The Institute of Biology's
Studies in Biology no. 72

The Dynamics of Competition and Predation

Michael P. Hassell

D.Phil.
Reader in Insect Ecology
Department of Zoology and Applied Entomology
Imperial College of Science and Technology
London

Edward Arnold

First published 1976
by Edward Arnold (Publishers) Ltd,
41 Bedford Square, London WC1B 3DQ
Reprinted 1980

Paper edition ISBN 0 7131 25853

Printed and bound in Great Britain at
The Camelot Press Ltd, Southampton

General Preface to the Series

It is no longer possible for one textbook to cover the whole field of Biology and to remain sufficiently up to date. At the same time teachers and students at school, college or university need to keep abreast of recent trends and know where the most significant developments are taking place.

To meet the need for this progressive approach the Institute of Biology has for some years sponsored this series of booklets dealing with subjects specially selected by a panel of editors. The enthusiastic acceptance of the series by teachers and students at school, college and university shows the usefulness of the books in providing a clear and up-to-date coverage of topics, particularly in areas of research and changing views.

Among features of the series are the attention given to methods, the inclusion of a selected list of books for further reading and, wherever possible, suggestions for practical work.

Readers' comments will be welcomed by the author or the Education Officer of the Institute.

1976

The Institute of Biology,
41 Queens Gate,
London, SW7 5HU

Preface

This booklet is intended to give an introduction to the ecology of competition and predation. Few species of animals do not suffer at the hands of predators and few animals or plants do not at times compete amongst themselves or with other species. Both processes, competition and predation, are important in affecting the size, stability and distribution of populations. On a different time scale, they result in the appearance of many evolutionary adaptations. It is not surprising, therefore, that much of the basic ecological theory that has arisen during this century is concerned with the importance of competition and predation.

The subject has direct relevance to man's own behaviour. We compete widely with other species, especially for space, and we are also a major predator, often in danger of over-exploiting our prey. From the ecological study of competition and predation we begin to understand the principles that underlie these interactions with vital resources.

I wish to thank Drs J. H. Lawton, V. C. Moran and D. J. Rogers who kindly read the text and made many helpful suggestions.

London 1976

M. P. H.

Contents

1 The Nature of Competition

1.1 Introduction

Population dynamics is the study of animal and plant populations; their patterns of change in space and time and the factors affecting these. Many factors, both physical and biotic, combine to determine how a particular population fluctuates in numbers. Some are important, in *causing* the observed fluctuations. Others act primarily to reduce the scale of fluctuations and to maintain them about an average, or equilibrium, population level—they are responsible for population regulation (SOLOMON, 1969). The study of these factors is a central theme of population ecology. In this book, we are concerned with competition and predation, the two most important biotic processes affecting a population's dynamics. All species, faced with limiting resources, will compete amongst themselves or with individuals of other species and few animal populations are free from predation. So competition and predation must be important factors affecting the evolution of species; natural selection will favour the superior competitor, the efficient predator and the elusive prey.

The complete study of these processes embraces several fields of biology. We should understand the mechanisms of competition and predation; what special prowess, for example, enables one individual to out-compete another? Amongst competing plants we should look for differences in growth form, vigour, reproductive rate or the ability to produce chemical inhibitors. Competition between animals similarly hinges on body size, vigour, reproductive rate and voracity, and also in certain groups on social status. We should also know something of the general ecology of competing species, predators and prey, if only to understand why they should be in the same place at the same time. Finally, we should certainly know how competition and predation affect the pattern of change in numbers of the interacting populations—to which this book is largely devoted.

1.2 What is competition?

Competition may be for food, space, mates or an elevated status within a social hierarchy. Fly larvae within dung pats, for example, may compete directly for the limited food available. Barnacles, on the other hand, compete for space on exposed rock surfaces. CONNELL (1961) has shown that individuals of *Chthamalus stellatus* are 'smothered, undercut, or crushed' by the faster growing individuals of *Balanus balanoides*. Surviving

Chthamalus tended to be smaller than average and therefore produced fewer progeny. Male fiddler crabs, too, compete for space. Here the competitive 'instrument' is the greatly enlarged and coloured claw of the left- or right-hand side. By display and combat, the fiddler crabs strive to secure and maintain a territory in intertidal zones, without which they cannot hope to secure a mate and reproduce.

The common elements in all these examples are that two or more individuals are *striving* against each other to secure some *resource* that is in *limited* supply. The 'limited resource' and 'strife', implying some harm to the weaker competitors, are essential ingredients for all competitive situations. Clearly, competition may be between individuals of the same species as in fiddler crabs (=intraspecific), or of different species as in the barnacle example (=interspecific); these are contrasted in Chapters 2 and 3. In this first chapter, we consider two very general questions: how may competition be manifest and how should these effects be measured?

1.3 Effects of competition

The effects of competition depend upon the resource which is in short supply, whether it be nutrients, space, available mates or some less definable commodity such as rank in a social hierarchy. Many insects provide excellent examples of animals competing directly for a food supply; we shall see in the next chapters how widely they have been used in laboratory competition experiments. The unsuccessful competitors may emigrate with the chance of finding richer pastures, or they may starve to death. Others, slightly more fortunate, survive but are reduced in body size. They may then suffer higher mortality during their later development and as adult females may have a lower fecundity (a reduced *natality*). These time delays between *cause* and *effect* are very important in population dynamics and are further discussed in Chapters 2 and 3.

Space is a limiting commodity to many animal and plant populations. Plants that are overcrowded may be denied their optimal amount of nutrients, water and light. Under these conditions their growth is stunted and they may perish. The maximum level of crowding is determined by the spatial or territorial requirements of the individual. For example, the maximum crowding of the bivalve mollusc, *Tellina tenuis*, on sandy shores is determined by the sweep of their exhalent siphon, and that of the aphid, *Drepanosiphon platanoides*, on sycamore leaves by the minimum spacing such that individuals rarely touch each other (KENNEDY and CRAWLEY, 1967). Many animals, such as the fiddler crabs and of course a wide range of bird species, depend upon space in the form of a territory for a secure supply of food or for attracting a mate. Competition is then most severe when the animals are more abundant than the available territories. Food and mates will also be linked where competition is for a higher rank, as found in many social groups particularly amongst mammals. The highest ranking animal has the prize of first pick of both food and females.

1.4   Detecting competition

Within the confines of a laboratory experiment, competition is often the clear cause of the replacement of one species by another (Chapter 3). However, competition between species under natural conditions is often deemed to have occurred by inference. When we see grey squirrels replacing red squirrels over much of England, it is tempting to think of the grey squirrel as competitively superior, especially since red squirrels still survive in their original habitat on Brownsea Island which grey squirrels have yet to invade. There is, however, little or no *direct* evidence that the two species compete at all. Food does not appear to be a limiting factor. Yet it cannot be denied that the decline of the red squirrel coincided with the increase of grey squirrels. It is unlikely that the two are unrelated. The apparent replacement of one species by another always suggests that competition is the driving force, but it is never sufficient evidence in itself. The effects of intraspecific competition may be more easily detected, especially when food or space are obviously limiting. But many populations are maintained below this limit and the effects of competition are difficult to detect by observation alone. Faced with this kind of uncertainty, we must consider what measurements are necessary to establish that competition has occurred.

A common feature of all competitive situations is that competition for a given amount of resource is fiercer, or more widespread when more individuals are present. In other words, the intensity of competition, expressed as the *proportion* of the total population that die, emigrate, fail to reproduce, etc., increases with the number of competitors. This is clear from the examples in Figs. 2–1 and 2–2. There are three essential ingredients required to express the effects of competition in this way.

(1) The number (or density per sampling unit) of competing individuals is a primary requisite.

(2) Competition will not be detected in the first place without knowing its effects on other competitors. We must, therefore, estimate the extent of the harm done to the weaker competitors (or the gain to the victors) and express this as the proportion of the total number of competitors that are affected—whether it be in terms of survival, reproduction or dispersal.

(3) The density of competitors in itself does not determine the level of competition. It is the population density *per unit of the resource* that is critical. Thus, apart from identifying the critical resource, it is important to know how much of it is available.

With these three estimates, the proportionate effects of competition may be plotted against population density per unit of resource. The use of percentages tends to yield curvilinear relationships that rise at first steeply and then more gradually towards an upper limit, which is usually the 100% mark. Such relationships tend to become more linear on logarithmic scales. For this reason, the examples of intraspecific competition in the next Chapter make particular use of *k-values* to express these proportionate effects. Originally due to HALDANE (1949) and widely used in the context of life table analysis by VARLEY and GRADWELL (1970), *k*-values are calculated from

$$k\text{-value} = \log_{10} X_t - \log_{10} X_s = \log_{10}(X_t/X_s)$$

where X_t is the initial population density and X_s the population density of survivors (in the case of mortality). Each percentage or proportion, therefore, has its corresponding *k*-value: 50% mortality, for example, is the same as a fractional survival of 0.5 which is always the same as a *k*-value of 0.301; 90% mortality is equivalent to a *k*-value of 1.0; 99% mortality to a *k*-value of 2.0 and so on.

The results of competition expressed in this way are revealed as *density dependent* relationships. The concept of density dependence is an important one going back to HOWARD and FISKE (1911) who coined the term 'facultative agencies' for any factors 'which affect the destruction of a greater proportionate number of individuals as . . . [they] increase in abundance'. SMITH (1935) replaced this with the term 'density dependence' which applies whenever the mortality or emigration rate increases, or the natality or immigration rate decreases, with rises in population density. Density dependent relationships must therefore tend to prevent the unlimited increase in population numbers and will usually contribute to population regulation. This is discussed in more detail in Chapter 2 for intraspecific competition. Interspecific competition can have fundamentally different effects on the dynamics of competing populations and is therefore treated separately in Chapter 3.

2 Competition within a Species

2.1 Scramble and contest

There are several ways in which a resource may be divided amongst competitors. NICHOLSON (1954) distinguished between two extremes of this, calling them *scramble* and *contest* competition. Pure scramble ideally involves the exactly equal partitioning of the resource and hence an equal division of the effects of competition between the competitors. This could be manifest by a sudden change, as population density increases, from complete survival to 100% mortality where there is just insufficient resource to maintain any individual. Examples of such scramble that approach this extreme are shown in Fig. 2–1a and b, where mortality of the blowfly (*Lucilia cuprina*) and fruitfly larvae (*Drosophila melanogaster*) respectively rises very rapidly above some threshold population density. Below this threshold, best seen in Fig. 2–1b, there is adequate food for the survival of most of the population. Alternatively, scramble may also be manifest by changes in body weight or in the number of eggs laid per female, rather than by mortality, in which case the effects will be spread equally amongst the competitors. Figure 2–1c shows an example of this, again from *Drosophila*, where the surviving adults show the effects of larval competition by a reduction in body weight. The tendency to scramble for food by the larvae was therefore having a marked effect at densities well below those at which larvae actually starved. Similar relationships may be expected from some annual plants. Agricultural crops, for instance, will 'scramble' for resources if too crowded; their yield per plant will decrease over a relatively wide range of increasing densities.

In general, such competition is typical of species that exploit rather transient food supplies such as carrion, dung or stored products. The individuals need to complete their life cycle as rapidly as possible before the food supply disappears or becomes unsuitable. There is therefore little opportunity for other forms of competition based, for example, on combat or the acquisition of a territory which would need constant defence. These kinds of competition tend to be characteristic of more stable habitats with more permanent resources, where the evolution of very rapid rates of population growth may not be an advantage. Instead, a species may well evolve more complex methods of ensuring a sufficient supply of resources for itself. Many of these fall within the category of *contest* competition.

Contest differs from scramble in that the resource is *unequally* partitioned; some individuals get all they require while others have insufficient for survival or reproduction. Contest would occur, for

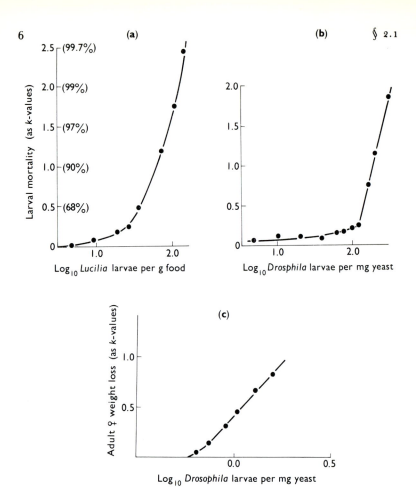

Fig. 2–1 Examples of scramble competition. (a) and (b) Mortality (expressed as k-values with percentages also shown in (a)) of *Lucilia* and *Drosophila* larvae competing for a fixed amount of food. (c) Reduction in female *Drosophila* weight at different larval densities. (*Lucilia* data from NICHOLSON, 1954. *Drosophila* data from BAKKER, K. (1961), *Archs. néerl. Zool.*, **14**, 200–81. After VARLEY *et al.*, 1973.)

example, where individuals compete for a fixed number of refuges (for survival), territories (for reproduction) or a position in a social hierarchy. In all these cases, the final maximum number that survive, breed, etc., will tend to be constant and thus independent of the numbers initially present. Contest can therefore be seen as a mechanism which will tend to maintain a constant population size as long as the amount of resource does not change. It very obviously promotes *stability*. A good example of this, shown in Fig. 2–2, is the effect of crowding on the survival of trout fry. Mortality is shown from curve (A) and the surviving population density from curve (B). Clearly, the maximum number of surviving trout is approximately 10 per m², irrespective of the initial stocking densities.

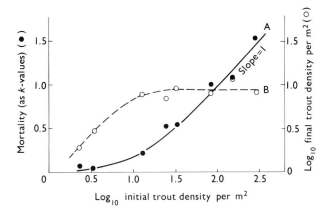

Fig. 2–2   The effects of competition in trout fry. Curve A: Mortality as a function of the initial stocking density of trout. Note that the relationship at high trout densities has a slope of unity. Curve B: The surviving trout density at different initial stocking densities. (Data from LE CREN, E. D. (1973), in: *The Mathematical Theory of the Dynamics of Biological Populations*, pp. 87–101. Academic Press.)

The essential characters of contest and scramble, as they affect mortality, are distinguished in Fig. 2–3a and b. In both cases there is no mortality below a threshold population density when there should be ample resource for all. Above this threshold, mortality rises abruptly to 100% in the case of perfect scramble (compare Fig. 2–3a and Figs. 2–1a and b), but much more gradually for perfect contest, in fact linearly with a slope of unity when k-values are plotted against the log density of competitors (compare Fig. 2–3b and Fig. 2–2A). This follows necessarily from the requirement that contest leaves a constant number of survivors, or victors, for a given amount of resource.

Almost all competition under natural conditions falls between these two idealized extremes. Abrupt changes as in Fig. 2–3 are not found in the real world; given the normal genetic and phenotypic variability present in populations, some individuals will always be 'fitter' than others. Thus, where a transient resource is being rapidly exploited, some individuals will always secure more of the resource and have a greater chance of survival. Similarly, pure contest is rare under natural conditions. Even where individuals compete for territories, there will often be a compromise as competition becomes more intense; average territory size may be reduced and more use made of marginal areas. The importance of the concepts lies mainly in defining the range of behaviour that should be incorporated in models of population growth. When this is done, the

8

§ 2.2

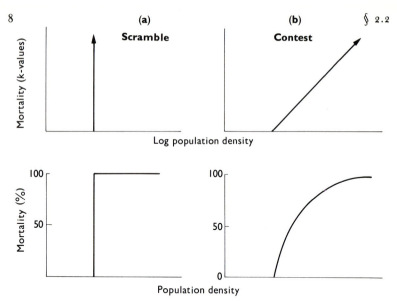

Fig. 2–3 The relationship between mortality and population density for pure scramble (a) and pure contest (b) shown on logarithmic and arithmetic scales.

range of *effects* that competition can have on populations may be explored.

2.2 Models for single-species competition

The models considered here aim solely to *describe* the net effects of competition on a population's rate of change in numbers. They can therefore be very simple and easily *used* to show how different levels of competition affect population fluctuations. This is one of the strategies of the mathematical modelling of ecological systems. MAY (1973) puts its objectives very well: the models 'sacrifice precision in an effort to grasp at general principles. Such general models even though they do not correspond in detail to any single real community, aim to provide a conceptual framework for the discussion of broad classes of phenomena.'

We commence with the simple statement that all populations have the potential to increase exponentially, as expressed by the equation

$$\frac{\mathrm{d}X}{\mathrm{d}t} = rX$$

$$(2.1a)$$

Thus the rate of change in the numbers of population X with the passage of time is the product of the numbers of X and their intrinsic rate of natural increase, r. This is the maximum instantaneous rate of increase under the given conditions of temperature, humidity, etc., as explained in

SOUTHWOOD (1966). To find the population size at any time t, we integrate to give

$$X_t = X_0 \ e^{rt} \qquad (2.1b)$$

where X_0 is the population size at time t_0, and from this we can plot the exponential growth of X with time as in Fig. 2-4A.

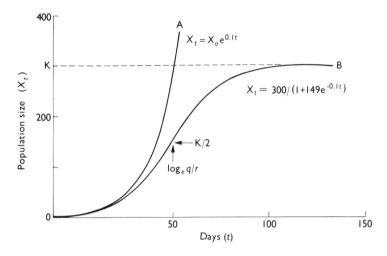

Fig. 2-4   Exponential (A) and logistic (B) population growth, showing the models for each and the point of inflexion on the logistic curve. (From VARLEY *et al.*, 1973; courtesy of Blackwell Scientific Publications.)

No population could sustain such increase for long. Competition for resources will become increasingly severe and the net rate of increase (dX/dt) therefore reduced, either due to increased mortality, reduced fecundity or both. VERHULST (1838), and later PEARL and REED (1920), expressed this idea mathematically in what is now called the *logistic equation*:

$$\frac{dX}{dt} = rX \left[\frac{K - X}{K} \right] \qquad (2.2a)$$

where K is the 'carrying capacity'—the maximum population size that can be sustained by a given amount of limiting resource. By integration, we have

$$X_t = \frac{K}{1 + q e^{-rt}} \qquad (2.2b)$$

where q is related to the point of inflexion on the time axis as shown in Fig. 2–4B. The population growth is now sigmoid. It commences almost exponentially but, as the population size increases, there is more and more feedback from the term $[(K-X)/K]$, representing the effects of competition. The net rate of increase therefore declines until, when the carrying capacity is reached $(X_t = K)$, there is no further change in population size $(dX/dt = 0)$ and the population is therefore at its equilibrium level $(X^* = K)$. The model is clearly a very stable one since the population will always tend to return to its equilibrium following a disturbance. Its essential character is one of contest competition; the maximum population size is independent of the initial density.

The logistic is a most simplified description of population change resulting from a complex biological process, and several of the assumptions upon which it is based are not biologically realistic. One of its important limitations is the assumption that everything acts instantaneously; there is an immediate effect of population size on subsequent population change via competition. In the real world some time delays between cause and effect are inevitable. They may arise by a population only reproducing at discrete intervals, in which case the effects of competition on natality are manifest only after the next breeding season. Time delays will also arise, for instance, where a population exploits a resource which is then depleted for subsequent populations. The delay is here associated with the resource recovery time.

Such delays can have marked effects on the stability of populations and should therefore be a feature of our models. They can be included explicitly, as in the following modification of the logistic (from HUTCHINSON, 1948)

$$\frac{dX}{dt} = rX_t \left[\frac{K - X_{t-T}}{K} \right] \qquad (2.3)$$

where T is the time delay. The feedback between competition and the rate of population change now depends upon population size at time $(t-T)$ rather than at t as in equation (2.2). Alternatively, the time delay may be *implicit* in the model by assuming that the generations are discrete. Now what happens in generation t is only manifest by its effect on generation $t+1$.

Consider, for example, the model (from HASSELL, 1975)

$$X_{t+1} = \lambda X_t [1 + aX_t]^{-b} \qquad (2.4)$$

where X_t and X_{t+1} are population sizes in successive generations and λ is the finite net rate of increase of the population per generation and hence directly related to the intrinsic rate of increase: $\lambda = e^r$ (see SOUTHWOOD, 1966). Such equations where time moves in discrete steps are known as difference equations. The constants a and b are best understood from Fig. 2–5. This displays the model by plotting the k-values for mortality

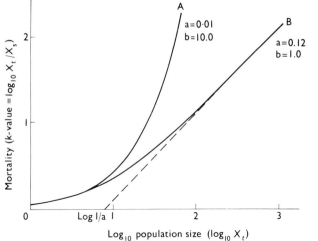

Fig. 2–5 Density dependent relationships from equation (2.5). Curve A approaches scramble, while curve B represents contest.

($\log X_t/X_s$) against log population size ($\log X_t$) (as in Figs. 2–1 to 2–3), where X_s are the numbers surviving competition that will go on to reproduce as defined by λ. The relationships are therefore described by

$$k = b \ \log_{10} (1 + aX_t) \qquad (2.5)$$

where b is the slope of the linear relationship attained at high population densities and $1/a$ the population density obtained by extrapolation back from this as shown in the figure. While the logistic equation (2.2) is limited to contest situations, equation (2.4) can describe the effect of competition from the extremes of scramble to contest, merely by varying the parameter, b, as done in Fig. 2–5. Curve (A) has the exponential character of scramble as in Fig. 2–3a, while curve (B) has that of contest as in Fig. 2–3b.

We are now in a position, by using these models, to explore the effects of these different types of competition on patterns of population change.

2.3 Effects on population change

Both the logistic model and equation (2.4) have the basic structure of a constant intrinsic rate of increase of the populations (r if instantaneous, or λ if discrete) and some density dependent feedback representing the effects of competition for a fixed amount of resource. Despite their great simplicity, they can show in a general way the kinds of effect that density dependent relationships have on populations.

In the absence of any density dependence, both models obviously reduce to simple models for exponential growth, provided that the growth rates fulfil the condition, $r > 0$ or $\lambda > 1$. The effect of competition is progressively to reduce the rate of increase as population size increases, which in most cases leads to an equilibrium position where successive populations remain of the same size. Having established the presence of an equilibrium, it is of particular interest to know how a population tends to approach this level. Does it do so exponentially? Can the increasing population overshoot the equilibrium with the associated danger of over-exploiting its resources? Can competition lead to population cycles? These questions are answered by investigating the *stability properties* of the models.

The essential stability properties of the logistic model have already been illustrated—in Fig. 2–4B. There is an equilibrium defined solely by the 'carrying capacity' ($K = X^*$) and the population is bound to approach this level exponentially. Such population behaviour has been observed where single species have been cultured in the laboratory with a fixed amount of food (Fig. 2–6). But so also have instances where the populations overshoot and oscillate markedly about some equilibrium, or average level (Fig. 2–7). The necessary conditions for oscillations are a high rate of increase (r or λ) of the population, strong competition and a sufficient time delay between a given population size and the subsequent effect of competition altering the population's rate of change as in the time delayed logistic (*2.3*) and the difference model (*2.4*).

To investigate this further, let us look more closely at the stability properties of equation (*2.4*). There are two parameters that affect stability, the competition parameter b and the rate of increase λ, while the parameter a only affects the size of the equilibrium population. Figure 2–8 shows examples where λ is fixed at a 20-fold increase per generation (a relatively high value) and a at 0.01, but b is varied between 0.5 and 5. With

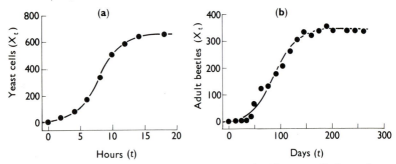

Fig. 2–6   Examples of sigmoid population growth. (a) Yeast cells (Data from PEARL, R. (1925), *The Biology of Population Growth*. Alfred Knopf, New York.) (b) Adult beetles (*Rhizopertha dominica*) per 10 g of wheat grains. (Data from CROMBIE, 1945.)

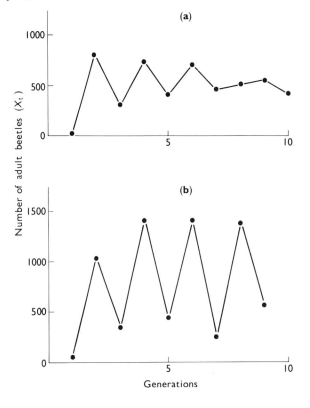

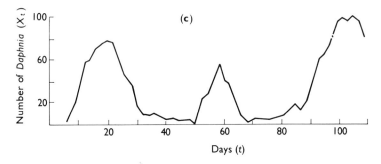

Fig. 2–7 Examples of single-species oscillations. (a) and (b) Two strains of the beetle, *Callosobruchus chinensis*, under identical culture. (Data from FUJII, K. (1968), *Res. Popul. Ecol.*, **10**, 87–98.) (c) The water flea, *Daphnia magna*, cultured at 25°C. (After PRATT, D. M. (1943), *Biol. Bull.*, **85**, 116–40.)

b less than unity (Fig. 2–8a), the population approaches its equilibrium value exponentially, much as in Fig. 2–4b for the logistic. Increasing values of *b*, however, lead to oscillatory behaviour; firstly to *damped oscillations* as in Fig. 2–8b and then to *stable limit cycles* where the population always tends to return to a cyclic configuration rather than to an equilibrium level (Fig. 2–8c shows a two-point cycle). Further increases in *b* lead to more complex limit cycles, which are repeated every four, then eight generations and so on. Indeed, when the values of *b* and λ are sufficiently large, the model has the property of producing irregular, 'chaotic' fluctuations which are quite aperiodic (MAY, 1975). These examples of population oscillations therefore arise by an alternation of very heavy mortality in one generation followed by reduced mortality and a rapid rate of increase in the next. Consider a population above its equilibrium level. There is now a heavy mortality due to the considerable effects of competition. In the next generation, therefore, the population is below the equilibrium. Competition is now much relaxed and, due to the high rate of increase λ, the population increases rapidly. In this way it tends to overshoot the equilibrium from side to side. With weaker competition ($b \leqslant 1$, Fig. 2–8a) or a much smaller potential rate of increase, oscillations will not occur. The return towards the equilibrium will now be more gradual, taking more than a single generation interval. This means that the rate at which the population approaches the equilibrium can be reduced in each successive generation. It is clear from this range of behaviour that we must be wary of the widespread notion that *all* density dependent factors contribute to the stability of populations. Very marked density dependence can itself be the cause of considerable population fluctuation.

The correspondence between Fig. 2–6 and Fig. 2–8a and that between Fig. 2–7a, b and Fig. 2–8b, c is striking. We are apparently able to account for the population changes in these simple laboratory experiments from the even simpler models of population growth in the face of limited resources. Low rates of increase and/or weak competition (some elements of contest alone) lead to smooth approaches to an equilibrium. Higher rates of increase and some elements of scramble lead to more rapid population changes and oscillations about the equilibrium. Populations that behave in this latter way are said to be *r*-selected (MACARTHUR, 1960), from the term in the logistic model. They tend to be opportunists dwelling in transient habitats (see page 5) with the need to increase rapidly before the habitat becomes unsuitable. There is therefore no great penalty in overshooting the equilibrium and over-exploiting their resource—they are mobile forms, well-adapted for seeking other similar habitats in which to reproduce (SOUTHWOOD, 1962). On the other hand, many species dwell in very stable habitats in which to over-exploit resources is disadvantageous to their progeny. These are said to be *K*-selected, also from the logistic. Their optimum strategy is to

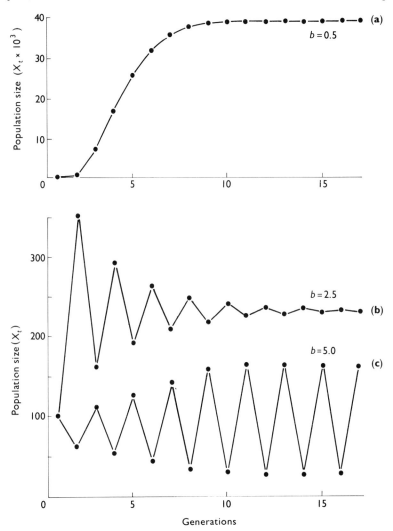

Fig. 2–8 Models of population change from equation (2.4) showing the effects of varying the parameter *b*, while *a* = 0.01 and λ = 20. (a) Exponential damping, (b) oscillatory damping and (c) two-point limit cycles.

remain as close as possible to their equilibrium. To rise above it is to over-exploit the resources; to fall below it is to give way to other species competing for the same resource (SOUTHWOOD, *et al.*, 1974). Such interspecific competition is an important factor shaping natural communities and is considered in the next chapter.

3 Competition between Species

3.1 Some experimental results

There have been many experiments, usually involving yeasts, Protozoa, various insects or plants, where two similar species are cultured together under laboratory conditions on a maintained food supply. Two types of outcome have been observed. Most commonly, the experiments lead to one or other of the species becoming extinct. This is well illustrated from PARK's (1948) experiments with two species of beetles, *Tribolium confusum* and *T. castaneum*, feeding on wholemeal flour to which 5% yeast had been added. At 29.5°C, Park found that *T. confusum* was normally replaced by *T. castaneum* as shown in Fig. 3–1a. Competition in such experiments is

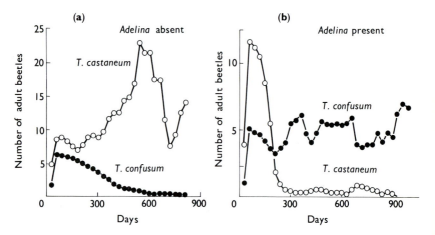

Fig. 3–1 Competition experiments between *Tribolium*. (a) The elimination of *T. confusum* by *T. castaneum* in the absence of the parasite, *Adelina*. (b) The reverse outcome when *Adelina* is present. (Data from PARK, 1948.)

often so delicately balanced between the two species that a slight change in the physical conditions can determine which species is to survive. Thus, PARK (1954), in further experiments with *Tribolium*, was able to reverse the outcome of competition merely by altering temperature or humidity. At temperatures above 29°C, *T. castaneum* tended to be the more successful competitor, but below 29°C *T. confusum* was favoured. Similarly, a change from 70% to 30% relative humidity at 34°C was sufficient to change the outcome from extinction of *T. confusum* to extinction of *T. castaneum*. A third important factor determining which species was replaced was the presence or absence of a protozoan parasite of the beetles, called *Adelina*. Park found that *T. castaneum* was the more susceptible species to *Adelina*, and that he could reverse the outcome shown in Fig. 3–1a to that in Fig. 3–1b solely by introducing the parasite to his cultures.

On the other hand, some experiments have led to the species coexisting. Figure 3–2, for example, shows the coexistence of *T. confusum*

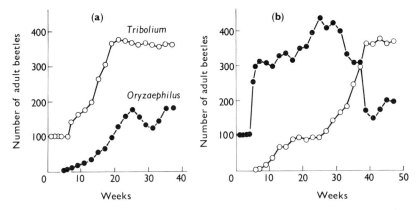

Fig. 3–2   Competition experiments between *T. confusum* and *Oryzaephilus*. Both species coexist and reach equilibria that are independent of the initial population sizes (compare (a) and (b)). (Data from CROMBIE, 1946.)

and *Oryzaephilus surinamensis*, another species of stored product beetle (CROMBIE, 1946). Both populations move to an approximate equilibrium position that seems independent of the relative densities at the start of the interaction. These experiments differed from those of Park by the inclusion of short lengths of glass tubing in the food medium, which served as refuges for the pupae of *Oryzaephilus* but were too narrow for *Tribolium*. The importance of such refuges to coexistence between competitors is discussed later in this Chapter.

Similar results have been obtained from greenhouse experiments with plants where various ratios of two plant species were established at the beginning of the growing season and compared with the ratio harvested at the end of the season (DE WIT, 1960). These experiments are well reviewed by KREBS (1972). They either indicate that one species is competitively superior to the other or that an equilibrium position between the two is likely.

Experiments such as these highlight a central problem in studying multispecies competition: to determine the general conditions that permit the coexistence of competing species. It is to this question that the basic theories are largely directed.

3.2   A theoretical background

The classical theory for two species competition is due to the pioneering works of LOTKA (1925) and VOLTERRA (1926). They start by assuming that each species (X and Y) on its own behaves according to the logistic model (equation 2.2). When together, they interact according to the equations

$$\frac{dX}{dt} = r_x X \left[\frac{K_x - X - \alpha Y}{K_x} \right] \qquad (3.1a)$$

$$\frac{dY}{dt} = r_y Y \left[\frac{K_y - Y - \beta X}{K_y} \right] \qquad (3.1b)$$

where α and β are the *competition coefficients* defining the equivalences between the species. Thus for species X, the rate of change dX/dt is zero, not when $X = K_x$ as in the logistic, but when $X + \alpha Y = K_x$, and similarly for species Y when $Y + \beta X = K_y$.

This model has the properties of either (1) species X or Y becoming extinct, when the model necessarily reduces to the logistic for the surviving species, or (2) both species coexisting together, each at their own equilibrium level. By exploring this further we can show the necessary conditions for these outcomes of competition which then help us to account for observed instances of coexistence or replacement.

A convenient means of illustrating the properties of any two-species model is to plot the numbers of species Y against species X at any point in time (said to be a plot in *phase space*) as shown in Fig. 3–3. In particular, we are here interested in the loci where each species just ceases to change in number: $dX/dt = 0$ and $dY/dt = 0$. These are the *zero growth isoclines*, as derived in the figure legend. Below its isocline, each species increases in number (e.g. $dX/dt > 0$), while above the isocline it decreases ($dX/dt < 0$). The possible outcomes of the Lotka–Volterra model depend on the relative positions of these isoclines and whether or not they intersect. The four alternatives are shown in Fig. 3–3a to d. If the isoclines do not intersect, no equilibrium between the two species is possible; the species whose isocline lies below the other inevitably moves to extinction (Fig. 3–3a, b). The system then behaves exactly according to the logistic model. We have seen that such replacement is common under experimental situations (e.g. Fig. 3–1). In biological terms, the species that survives is the one having the greater effect on the other species—the dominant competitor. Figure 3–3c and d differ in having *competitive equilibria* where the isoclines intersect. At this point, both populations are perfectly stable. That in Fig. 3–3c, however, is an *unstable equilibrium* point; the slightest disturbance leads to the elimination of one of the species depending on the direction of the disturbance. The necessary condition for this, that $K_y > K_x/\alpha$ and $K_x > K_y/\beta$ corresponds to the biological

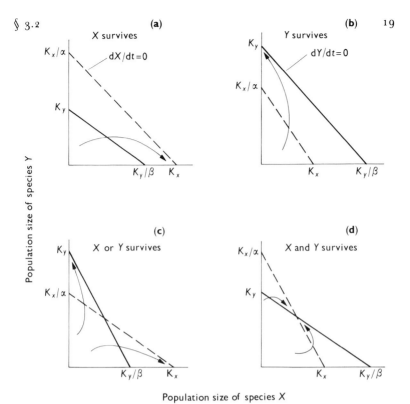

Fig. 3-3   The four possible outcomes from the Lotka–Volterra model (*3.1*). The zero growth isoclines are shown (broken line for species *X*) and arrows indicate which species survive. The intercepts on the axes that define the zero isoclines are obtained as follows, taking species *X* as an example. When $dX/dt=0$, then $K_x-X-aY=0$. Thus, $Y=K_x/a$ when $X=0$, and $X=K_x$ when $Y=0$.

situation where each species inhibits the population growth of the other species more than itself. In other words, *interspecific* competition is much stronger than intraspecific effects. Finally, a *stable equilibrium* point is possible (Fig. 3-3d) if $K_y < K_x/a$ and $K_x < K_y/\beta$. Each species must now inhibit its *own* growth more than that of the other species and hence *intraspecific* effects must outweigh interspecific ones.

The Lotka–Volterra model is very obviously derived from the logistic model for single species growth and therefore retains many of its properties. In particular, each population can only approach its equilibrium exponentially. Introduce time-delays, however, and the range of stability behaviour increases to include the kinds of oscillatory behaviour discussed in Chapter 2. Consider, for example, the single-species model of equation (*2.4*), namely

$$X_{t+1} = \lambda X_t [1 + aX_t]^{-b}$$

This is easily extended to the equivalent two-species model by introducing the same interspecific competition coefficients (α and β) as previously:

$$X_{t+1} = \lambda_x X_t [1 + a_x(X_t + \alpha Y_t)]^{-bx}$$

$$Y_{t+1} = \lambda_y Y_t [1 + a_y(Y_t + \beta X_t)]^{-by} \qquad (3.2)$$

This model has very similar requirements to the Lotka–Volterra model for the coexistence or replacement of the two species, depending again on how the linear isoclines ($X_{t+1} - X_t = 0$ and $Y_{t+1} - Y_t = 0$) intersect. The real difference is that the populations can now oscillate about the equilibrium as shown in Fig. 2–8 for the single-species case. Whether they do so or approach the equilibrium exponentially, depends in a complex way on λ, α, β and b. The few experiments in which two species have coexisted do not show clear oscillations about the equilibria, although some tendency to overshoot the equilibrium is apparent from Fig. 3–2.

While the extension from single- to two-species models has no qualitative effect on the range of possible stability 'behaviour' about an equilibrium, it can have a marked effect on the very existence of an equilibrium for one or other of the species. We have seen that the single-species models in Chapter 2 will always have a positive equilibrium if $r > 0$ or $\lambda > 1$. This is due to the density-dependent intraspecific competition. Interspecific competition, however, is a destabilizing process. Coexistence between the two species is therefore only possible if the intraspecific effects outweigh the interspecific ones (see Fig. 3–3d). For example, if α and β in either of equations (3.1) or (3.2) are zero, the two populations remain quite independent. As α and β increase in value, so interspecific competition becomes more important and the possibilities of coexistence diminish. Finally, when the product $\alpha\beta > 1$, one or other species inevitably moves to extinction.

There have been few attempts to test any specific competition model experimentally. It is not sufficient merely to allow two species to interact and then to record the outcome. To test the Lotka–Volterra model, for example, estimates are needed of the equilibrium population sizes of each species on its own (K_x and K_y) and of the competition coefficients (α and β). Graphs such as Fig. 3–3 may then be drawn and used to predict whether or not the species should coexist and whether they do so at the predicted point. If possible the growth rate (r_x and r_y) should also be independently estimated in order to predict the pattern, or trajectory, of population growth of the two species during the course of the experiment. Such data were obtained by CROMBIE (1945, 1946) working on competition between various stored product beetles and moths, and allowed him to plot the zero growth isoclines for each of his experimental systems. His results provide good support for the model (Fig. 3–4a, b). Coexistence or extinction of one species occurred as predicted with the populations tending to follow the predicted trajectories. In contrast to

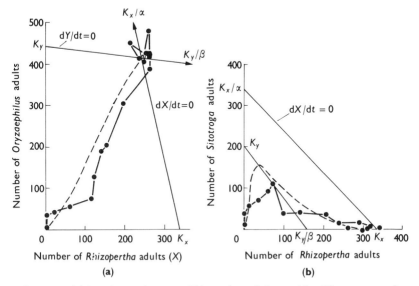

Fig. 3–4 (a) Coexistence between *Rhizopertha* and *Oryzaephilus*. The zero growth isoclines are shown and also the observed (●) and calculated (– – –) population growth curves. (b) A similar plot showing the replacement of the moth *Sitotroga cerealella* when in competition with *Rhizopertha*. (Data from CROMBIE, 1945.)

this, GILPIN and AYALA (1973) present some results of competition between two species of fruit flies (*Drosophila*), where the zero isoclines are now markedly concave and can only be predicted from more complex models; for instance, where competition is allowed between more than one age-class of the two competing species. Such complexities, and many more, will always occur under natural conditions.

The real value of the simple models is in emphasizing the general destabilizing effects of interspecific competition and in showing the essential conditions for coexistence of competitors. These are embodied in the Competitive Exclusion Principle.

3.3   The Competitive Exclusion Principle

Equations (3.1) and (3.2) only permit coexistence when the growth rate of each species is inhibited more by intraspecific than interspecific effects (Fig. 3–3d). For this to occur, the two species must have some difference in their needs, such that at some stage in their lives they are at least partially protected from interspecific competition. Conversely, two species cannot live together if they share exactly the same ecological requirements in time and space. This reasoning forms the basis of the Competitive Exclusion Principle, whose origins are well discussed by HARDIN (1960). He gives the straightforward definition that 'Complete competitors cannot coexist'.

This ecological 'theorem' has provoked considerable interest. It has, however, not proved easy to validate from experiment or direct observation. The fact that most laboratory competition experiments have led to the extinction of one of the species is certainly in accord with Competitive Exclusion since they afford little opportunity for avoiding continuous competition. Where coexistence has occurred, some refuge has usually been identified. Thus CROMBIE (1946) found from his experiments with stored product insects that only those species with slightly different ecologies survived together, while of those with exactly the same requirements one always eliminated the other. The coexistence between *Rhizopertha* and *Oryzaephilus* shown in Fig. 3–4, for example, results from a partial separation of feeding habits in the two species. Larval *Rhizopertha* feed within the whole wheat grains, while larval and adult *Oryzaephilus* and adult *Rhizopertha* all feed from the surface of the grains.

It is when we turn to species competing under natural conditions that the Principle becomes most difficult to apply, for now it is barely feasible that any two species could share identical *niches*. The niche here defines the sum of all ecological factors affecting a particular population; it is an n-dimensional hyper-volume where each of the n axes represents the range of a different ecological factor (e.g. temperature, moisture, light, food size, etc.) within which the species can survive and multiply (HUTCHINSON, 1958). It will always, therefore, be difficult, when two very similar species are seen to coexist, to distinguish between the Exclusion Principle refuted or some significant ecological difference overlooked. Probably the closest we can get to observing Competitive Exclusion in operation comes from some examples of species replacement after very similar species have invaded the same area. Figure 3–5, for example, shows the percentage parasitism by three species of insect parasitoids (see Chapter 4) in the genus *Opius* that were imported into Hawaii for the control of the Oriental fruit-fly (*Dacus dorsalis*). The three species of *Opius* were entirely dependent on the *Dacus* to complete their life cycles. *Opius longicaudatus* was the first established, but declined as *O. vandenboschi* increased in numbers. Both species were then eliminated when the third species, *O. oophilus*, increased in numbers. Perhaps the decline of the red squirrel in favour of the grey squirrel in England, mentioned in Chapter 1, is a further example.

Despite difficulties in employing the Competitive Exclusion Principle,

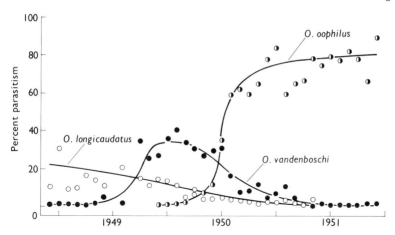

Fig. 3–5   Changes in percent parasitism of the fruit-fly *Dacus dorsalis*, by three species of *Opius* parasitoids. (Data from BESS, H. A., VAN DEN BOSCH, R. and HARAMOTO, F. H. (1961), *Proc. Hawaii. ent. Soc.*, **17**, 367–78. After VARLEY *et al.*, 1973; courtesy of Blackwell Scientific Publications.)

there is no doubt that species with the most similar niches will tend to compete the most intensively. But rather than look for consequent extinction, we should seek evidence for competition leading to changes in the species' morphology, behaviour or habitat, thus reducing the amount of niche overlap. There is a wealth of evidence that this occurs, mainly from studies of closely related bird species. In a classic study, MACARTHUR (1958) made detailed observations of five species of warbler in the genus *Dendroiea* in New England forests, all of which are insect feeders on the same species of spruce. He showed that the 'overlap' between the bird species is reduced by different feeding zones in the canopy, different movement patterns (of hovering, running and hopping) and slightly different nesting times. In a comparable study, LACK (1945) showed that the very similar cormorant (*Phalacrocorax carbo*) and the shag (*P. aristotelis*) have significantly different nesting requirements and tend to eat different food types. It is not difficult to accept that competitive pressures have led to these species specializing in different ways.

Further evidence comes from examples of *character displacement* (BROWN and WILSON, 1956), a term describing any differences in morphology or ecology that are *only* seen where the ranges of the species overlap. A fine example of this comes from two species of Galapagos finches, *Geospiza fortis* and *G. fuliginosa*, studied by LACK (1947). He found that where the two species coexist on the *same* island, there is no overlap in the range of their beak sizes. However, on some smaller islands where only one or other species occurs, the range of beak sizes coincides to a large extent. The inference here is clear; competition on the larger islands containing both species has led to each specializing on different food sizes and hence increased their ecological isolation.

3.4   Competition, niche overlap and species diversity

We have seen that the Competitive Exclusion Principle does not permit the stable coexistence of competing species whose niches completely overlap, and that competition generally results in some reduction of niche overlap between the competing species. A more useful statement of the Exclusion Principle would specify just how similar competing species can be and still persist together in a community, a problem to which much effort has been devoted (e.g. MACARTHUR, 1970; MAY, 1973, 1974). We here focus on a single study, that of MAY (1973, 1974), which provides a relatively simple basis for a general theory of niche overlap between competing species.

Let us consider the case where several species depend upon a single resource (such as food) which is present in a continuous spectrum of (say) size. This is schematically shown in Fig. 3–6 for three species each

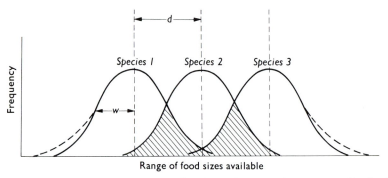

Fig. 3–6   Frequency distributions for the utilization of a resource (food of varying size) by three species. d represents the separation of the curves and w their spread. (After MAY, 1973; courtesy of Princeton University Press.)

preferring different food sizes but with considerable overlap in the food sizes taken (hatched area in Fig. 3–6). These frequency curves are assumed to be identical for each species and normally distributed about their means (although skewed distributions would be quite likely under natural conditions, especially at either end of the range (dotted lines in Fig. 3–6) where there is no pressure from competitors). They are characterized by two parameters alone; the separation of the species (d) along the resource spectrum and the spread of each distribution (w) equal to one standard deviation, as shown in the figure.

The problem is now not whether identical competitors can coexist, but how much overlap there can be between the species without extinctions.

By exploring this situation mathematically some general conclusions are reached on the limits to niche overlap consistent with a stable equilibrium between the species (i.e. allowing their persistence).

(1) In a deterministic environment, where no random fluctuations in the availability of the resource are allowed, there is no limit to niche overlap consistent with a stable equilibrium, short of complete overlap $(d=0)$ when coexistence suddenly becomes impossible. This, therefore, harks back to the Competitive Exclusion Principle where there must be some difference $(d>0)$ for coexistence.

(2) The conclusion is qualitatively different as soon as *any* random fluctuations are permitted. There is now a definite limit to the degree of overlap that still allows all species to coexist. If the random fluctuations are severe, the limit depends mainly on the variance of the fluctuations. But for all more modest levels of random variation the limit occurs when the separation of the species (d) is approximately equal to the spread of their distributions (w), or when $d/w \sim 1$.

These conclusions are robust in that they do not depend critically upon the specific assumptions in the model. Thus the shape of the frequency distributions may be changed and the spacing of the species made unequal $(d$ no longer constant), but the limits to niche overlap still remain that $d/w \sim 1$, provided that the random fluctuations in the resource availability are not very great.

Most species in the real world, of course, compete for several critical resources, some of which may be inter-dependent. May has investigated the extremes of this; where competition for each resource is quite independent of the others and where competition affects several resources in a complex fashion. He showed that the condition $d/w \sim 1$ remains true in the first case of independent resources, but where competition affects several resources at once, the limit to niche overlap tends to become proportional to the environmental variance once again. The further complication of predators affecting the coexistence of competing prey is considered in Chapter 7. These first steps in studying the dynamics of competing communities show that useful insights are to be gained from simple models, even when dealing with such complex systems.

Closely related to the problems of niche-overlap are those of *species packing*. How many species can coexist together along a resource spectrum? The models have so far avoided this question; the condition $d/w \sim 1$ could be attained equally well by very close species packing and narrow ranges of (say) food size taken (small d and w respectively), or by few species each with a broad spectrum of possible food sizes. There is as yet no simple theoretical framework to accommodate this general problem of what determines the level of species packing. This lack is partly compensated by a wealth of field studies on species diversity, even of whole phyla within geographic areas. On this global scale, the most intriguing problem is to account for the pronounced tendency for species diversity to increase towards the tropics, best seen from a comparison of a tropical rainforest with its temperate equivalent. Although many facets of this subject are outside the scope of this book (see KREBS (1972) for a good review), competition plays an important part and deserves mention.

A fundamental difference between tropical and temperate forests lies in their climate. Of particular importance is the higher average temperatures and equable climate of the tropical forest that not only leads to a higher energy (or biomass) turnover, but also a more seasonally predictable distribution of resources. This alone will enable species to survive in more restricted niches and permit greater specialization. Certainly, tropical rainforests abound in specialists of all kinds, showing refined adaptations in behaviour, utilization of space and also of time (breeding at different times of year, for instance). Species in the less equable temperate forests would not in general survive in such restricted niches; they must have a larger effective niche volume. Thus, the very wide range of exclusively fruit-eating birds and bats and birds specializing on large insects that are found in the tropics could not be supported in a temperate forest where fruit production is so seasonal and large insects are not always present.

The greater physical division of a tropical forest is well illustrated by MACARTHUR and MACARTHUR (1961). They found that the differences in bird species diversity in a tropical forest in Panama compared with a North American forest could be largely explained by the greater vertical stratification of canopy volume and also because the tropical birds tended to subdivide this resource dimension more finely than their temperate counterparts.

While greater resource predictability will certainly permit greater specialization and species packing in the tropics, the essential ingredients so far lacking are the underlying selection pressures forcing species to be so restricted. It is here that competition is of great importance. We have already seen from the examples on page 24 that competition can lead to increased specialization. The reasons for this competition being more widespread and intense in the tropical rainforests compared with temperate woodlands are not hard to find. The rainforest has relatively few abiotic selection pressures shaping the life history strategies of species. The climate is equable and resources tend to be in continuous supply. Under these conditions, competition is an important selection pressure leading to many of the specializations of the tropics. Temperate regions, on the other hand, have a much harsher climate to which species must adapt in order to survive. In particular, there is a considerable premium for suitable life styles for the cold winter months. Competition is therefore less intense and only one of many selection pressures that shape the composition of temperate communities.

4 Predator–Prey Interactions

4.1 Introduction

An interaction which is very different from competition is one in which one species feeds directly upon another. This includes herbivory, predation and parasitism. The term *predation* is usually restricted to the killing and eating of animals (the prey) and is, of course, a widespread habit in the animal kingdom. Predatory plants are much rarer. There are approximately 500 species, mainly of pitcher plants, sundews, flytraps and bladderworts, all of which secure insects or other small invertebrates by a pitfall, sticky hairs or a sprung mechanism. The prey are then digested and the products, particularly nitrogenous ones, absorbed. Animals show a much greater range of strategies for securing prey. There are filter-feeding predators (blue whales), predators that trap their prey (web-spinning spiders) and others that rely mainly on ambush (leopards and birds of prey), stealth (chameleons), speed (cheetahs) and so on.

Predation, like competition, is an important factor in the evolution of new species characteristics by natural selection. Selection will tend to favour those adaptations that decrease the vulnerability of prey, on the one hand, and increase the efficiency of predators, on the other. The wealth of examples where animals show cryptic, aposematic or mimetic colouration is itself testimony to the general importance of predation as a factor reducing survival. In general, only those species that are very large (elephants) or are themselves at the top of their food web (lions) are free from predation, except at the hands of man.

The principal problems in studying the dynamics of predation are to explain the observed distribution and changes in abundance of predator and prey populations. To do this we must firstly be aware of some general features of predators and predator–prey interactions. This chapter therefore includes a discussion on (1) insect parasitoids—a rather special kind of predator, (2) the range of prey species attacked by one predator species and (3) the relative synchrony of predator and prey life cycles and the importance of prey refuges. We then look at some general components of predator searching behaviour that affect the prey death rate and also the factors affecting the rate of increase of predator populations (Ch. 5). This is followed by a review of some simple population models (Ch. 6), and how these relate to natural populations (Ch. 7).

4.2 Insect parasitoids

Most predators are easily identified as such and come from all parts of the animal kingdom; from Protozoa to mammals. There is, however, one very large category of insects of great ecological importance whose predatory habit is not so obvious. These are the insect *parasitoids*, often loosely called 'insect parasites'. They largely belong to two orders, the Diptera (flies) and Hymenoptera (ants, bees, wasps), and differ from true parasites (like tapeworms) in the strict zoological sense in that they almost always kill their hosts. The hosts may belong to almost any of the insect orders (some other arthropods such as spiders and woodlice are also attacked), and egg, larval and pupal stages are more often attacked than adults. Insect parasitoids are very abundant, making up approximately 10% of all the one million or so known insect species. They play an important part in the natural control of other insect populations and this has led to their wide use in the biological control of insect pests (see Chapter 7).

The parasitoid life cycle is in some important respects much simpler than that of most other predators. This makes them ideal experimental animals for investigating some aspects of predation (see Chapter 5) and also means that they apply particularly well to the theoretical work discussed in Chapter 6. The hosts are normally only located by the adult female parasitoid who lays her eggs on, in or near the host individuals. The behaviour of the adult females varies considerably between species and can be most sophisticated. For example, some species can locate their hosts olfactorily, either responding to volatile substances from the host food-plant or from the host themselves (see section 5.4); some species paralyse their hosts before oviposition; some feed on host fluids to get adequate protein to maturate their full egg complement and some are efficient at detecting whether or not a host has already been parasitized. There is considerable variation in the number of eggs laid for each host found. A single egg is laid where only a single parasitoid larva can develop and mature from one host. If the same host is parasitized on separate occasions, it is usual for the first parasitoid larva present to be the sole survivor. On the other hand, other parasitoids lay several eggs per host, or sometimes a single egg which divides to give rise to several larvae (polyembryony), all of which can complete their development from the one host. When the parasitoid eggs hatch, the larvae may feed from the outside of the host (as ectoparasitoids) or from within (as endoparasitoids). Initially, they cause little serious damage; they feed as true parasites. However, as the parasitoid larvae approach maturity, they begin to feed on vital host organs and usually the host is killed by the time the parasitoid is ready to pupate. Pupation may occur close to or within the remains of the host.

There are, therefore, several significant differences between insect parasitism and predation.

(1) The death of the host following 'capture' is delayed until the parasitoid larvae are fully developed. This has little effect on the dynamics of the

interacting populations except in the very few cases where an adult host can produce a few progeny before being killed. We shall see that in the theoretical studies discussed in Chapter 6, the host is effectively assumed to be killed when first attacked and the interval between oviposition and emergence of the subsequent adult parasitoid is the generation interval.

(2) Only the adult female parasitoid searches for hosts. This is an important simplification. It means that only one set of parameters describing the outcome of parasitoid search need be included in a population model. In contrast, the males, females and often the immature stages of predators search for prey. They are likely to have different searching abilities (see Chapter 5) and may search for completely different prey types. Adult dragonflies, for example, feed on small insects taken in flight, while the nymphs are voracious aquatic predators. The size of prey taken will depend on the predator's stage of development.

(3) The number of hosts attacked by a parasitoid population *must* closely define the number of progeny in the next parasitoid generation, since each attacked host gives rise to a relatively constant number of parasitoids in the subsequent generation. There will be a one-to-one relationship in the simplest cases where only a single parasitoid can develop from each host parasitized. Predator reproduction, on the other hand, is more difficult to define. Some factors affecting the fecundity of predators are discussed in Chapter 5.

4.3 Predator specificity

Predators (including parasitoids) may be *specific* or *monophagous* (attacking a single prey species), *oligophagous* (attacking a narrow range of prey, usually from one family) or *polyphagous* (attacking a much wider range of prey). Completely specific predators are few. They have sacrificed a broad spectrum of food types for the relatively short-term advantages of very narrow specialization and, of course, they are entirely dependent on their single prey's persistence. On the other hand, to accept the widest possible range of prey species is also unlikely to be an optimum strategy. Instead, it may well be advantageous for the 'prudent' predator to specialize on those prey species which are relatively more profitable in terms of *net* energy gain per item eaten (i.e. the difference between the energy gained from eating one prey and that expended in its capture). There will be a tendency towards specificity when there is a single prey species that is markedly the most profitable, and towards polyphagy when several prey species all present more-or-less the same profitability. The optimum number of prey species in a predator's diet should therefore occur when the inclusion of yet a *further* prey species would lower the average profitability over all the prey (MACARTHUR and PIANKA, 1966). Of course, net energy gain per item eaten is not the sole factor determining these predator strategies. Prey profitability must always be balanced against prey availability, for example. For specific predators, in particular, it is essential that the prey are regularly available for the searching predators. Some aspects of this are discussed in the next section (4.4). Theoretical population studies have, for the sake of simplicity, largely concentrated on single predator–single prey systems (Chapter 6). There is now, however, a growing interest in the dynamics of polyphagous predator–prey interactions, particularly in the ways that different potential prey populations are exploited as their relative abundances change. It is common for polyphagous predators to show a preference for some prey over others, depending upon such factors as the size and palatability of the prey and their ease of capture. Prey that are very difficult to catch or yield little suitable food will tend to make up the diet of hard times. In addition to such innate preferences, the proportion of a particular prey type attacked will also depend upon its abundance relative to other kinds of prey. Many predators, particularly amongst the vertebrates, can learn to concentrate their search for the most abundant prey. In a classic study, TINBERGEN (1960) found that the proportion of different prey species (mainly caterpillars) in the diet of nestling great tits varied with the relative prey densities during the breeding season. The proportion of a particular prey in the diet tended to be low when it was rare, increased rapidly as a prey's abundance increased and then levelled off with further increases in abundance, as shown by the hypothetical example, A in Fig. 4–1. Tinbergen invoked a learning process to account for this. He firstly assumed that the birds searched randomly, so that the number of encounters with a particular prey should depend upon its abundance. This assumption alone yields the linear relationship, B in Fig. 4–1. To explain the differences between A and B, Tinbergen suggested that a prey species tends to be overlooked by the searching birds when only available in low numbers so that the proportion taken will be less than expected on a random basis (a to b in Fig. 4–1). As the number of this prey increases, so the number of encounters with them will increase. This in turn would lead to the birds learning to concentrate on certain characteristic cues from the prey (size, shape, colour, etc.), which Tinbergen called the development of a *specific searching image*. It means that the birds become deft at picking out a particular prey that is abundant and therefore take a larger proportion than expected from random predation (b to c in Fig. 4–1). When the prey become less numerous (e.g. caterpillars pupating), the birds abandon this specific searching image in favour of another. The levelling off of curve A at very high prey densities was explained by the need to have a varied diet.

While the ability to concentrate on certain visual cues to the neglect of others is probably widespread in birds and other higher animals, it is not a complete explanation for all cases of switching from one prey type to another. We shall see in the next chapter that random search is more

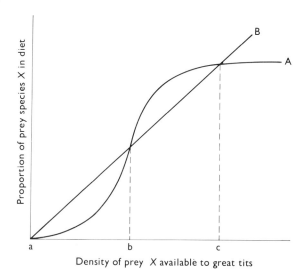

Fig. 4–1 Diagram illustrating the general response of great tits to changing densities of a particular prey species (X), as found by TINBERGEN (1960) (Curve A). B shows the relationship expected from random search and without searching images being formed. Points, a, b and c are explained in the text.

likely to be the exception rather than the rule and that most predators will tend to spend more time in those localities where prey are more numerous. For example, GIBB (1958) showed that the proportion of pine cones opened by great tits and blue tits hunting for larvae of the moth, *Ernarmonia conicolana*, tended to be higher in areas where there was a greater caterpillar infestation. The implication here is that the birds spend much less time in areas of low caterpillar densities than where their 'reward rate' is higher. This idea has been considerably developed by ROYAMA (1970), again studying the diet of nestling great tits. He suggested that Tinbergen's observations may be explained by assuming the different prey to be in slightly different habitats and the birds tending to spend most time in the most profitable areas.

4.4 Coincidence in time and space

Few predator populations are completely coincident with their prey. Their life cycles may be somewhat out of phase or some prey in particular locations may be free from predation. The degree of temporal or spatial coincidence is obviously an important factor in the dynamics of predator-prey interactions. It is most important to monophagous and oligo-phagous predators. Polyphagous predators, such as shrews (*Sorex*) that feed widely on arthropods, worms and other invertebrates (DELANEY,

1974), are unlikely to be (and need not be) coincident with any given prey species.

Temporal coincidence, or *synchrony*, clearly implies 'being together in time'. Selection will favour predators that are well synchronized with their prey, but at the same time will favour prey that can avoid this. Such a definition is useful in a general sense. More precise definitions have tended to focus on particular aspects of synchrony. GRIFFITHS (1969), for example, proposed an index of synchrony, I_s, where

$$I_s = T_c / T_{pred} \qquad (4.1)$$

T_c being the number of days that prey and predator are in contact and T_{pred} the predator life span. Figure 4–2a shows an example of perfect

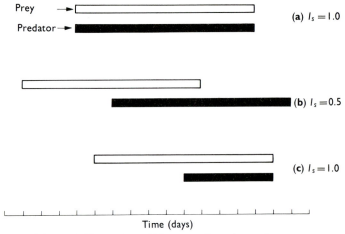

Prey
Predator

(a) $I_s = 1.0$

(b) $I_s = 0.5$

(c) $I_s = 1.0$

Time (days)

Fig. 4–2 Schematic illustration of degrees of synchrony between a prey and predator population, showing the indices of synchrony from equation (4.1).

synchrony $(I_s = 1)$ and Fig. 4–2b a case of partial synchrony $(I_s = 0.5)$. This definition takes very much the predator's view point: perfect synchrony requires each predator to be always exposed to prey without demanding the prey to be always exposed to predation. In Fig. 4–2c, for example, $I_s = 1$, but the prey are free from attack for half their period of availability. This would alter the dynamics of the interaction considerably (see Chapter 6). Changing the above equation to

$$I_s = T_c / T_{prey} \qquad (4.2)$$

where T_{prey} is the duration that prey are available, would now make the index sensitive to any period when the prey are not exposed to predation rather than when the predators have no available prey.

The observed range of synchrony between predator and prey populations is great. Complete synchrony is approached in monophagous species (particularly parasitoids) where their life cycles have evolved to fit that of their prey. Polyphagous species, on the other hand, may be poorly synchronized with any given prey species, tending to be opportunists and feeding on whatever prey is abundant at the time.

In the same way as prey may escape predation due to asynchrony between the searching populations, there may be *refuges* for prey due to imperfect spatial coincidence. Such refuges will tend to protect either a fixed *proportion* or a fixed *number* of the prey population at any one time. Flour moth caterpillars (*Ephestia*), for example, are protected from the parasitoid wasp, *Nemeritis*, when they are sufficiently deep in the flour to be out of reach of its ovipositor. In this way only a proportion of the host's habitat is searched. Clown fish, on the other hand, are protected from predators only when within the tentacles of an anemone species. This provides only a fixed number of refuges depending upon the size of the anemone population. In many cases, of course, the refuges will be less precisely defined. For example, cabbage aphids (*Brevicoryne brassicae*) form large, tightly-packed colonies on cabbage and other Brassica plants. There is a tendency for those individuals at the periphery of the colonies to be more susceptible to parasitism by the wasp (*Diaeretiella rapae*) than those towards the centre. This refuge effect will therefore increase with the area of the colony.

An index of spatial coincidence, I_c, useful for such examples was given by GRIFFITHS (1969):

$$I_c = N_0/D \qquad (4.3)$$

where I_c is the proportion of susceptible prey, N_0 is the number of prey susceptible to attack and D is the total prey population. This index becomes unity when no prey are completely protected from attack and tends to zero as more and more of the prey are in refuges. The effects of such refuges on the dynamics of an interaction are discussed in Chapter 6.

In this chapter we have seen a little of the range of behaviour and types of interaction with prey shown by predators. A primary concern for the population ecologist is to understand how these varied aspects of predation affect the interacting populations. Are they important to the stability of the population; do they mainly affect the equilibrium levels, or are they relatively unimportant to the overall dynamics? These are all questions most easily answered by the development of simple theoretical models. It is not intended that the models should represent every aspect of predation. To do so would lead to a cumbersome theoretical framework within which the effects of important factors would be difficult to distinguish. Much theoretical work has, therefore, centred around the effects of specific and synchronized predators in population interactions and, in particular, on the factors affecting the searching performance of a predator population. These will now be discussed in some detail in the following two chapters.

5 The Components of Predation

5.1 Some essential components

The first theoretical models of predator–prey interactions were developed about half a century ago by LOTKA (1925), VOLTERRA (1926) and THOMPSON (1924), soon followed by that of NICHOLSON and BAILEY (1935). These early attempts, discussed in Chapter 6, were purely deductive: there was no attempt to verify assumptions from field or laboratory data. More recently this trend has been reversed. There are now considerable experimental data available which can be used to determine the validity of earlier models and, very importantly, to help in developing more realistic general models. These models are not designed to describe particular predator–prey interactions under natural conditions. The usual objective is to model the simplest of interactions, concentrating only on representing in a realistic way the basic components of search and reproduction by a predator population. The effects of each component on the stability and equilibrium levels of the interacting populations may then be studied. There is, therefore, no emphasis on including the many other factors which affect both populations, whether they be due to climate, disease, parasites or other predators, and which would only tend to obscure the essential part played by each component of predation.

In this chapter, we shall distinguish between two aspects of predation.

(1) The death rate of the prey due to predation and, in particular, how the outcome of search by predators is affected by the abundance of prey (section 5.2), the abundance of other predators (section 5.3) and the relative prey and predator distributions (section 5.4).

(2) The rate of increase of the predator population (section 5.5), which must hinge on the survival of the immature stages as well as on the fecundity of the female predators.

The general form of these different relationships is now fairly well understood due to the wealth of experimental studies, mainly from arthropod predators and parasitoids (see BEDDINGTON et al. (1976), HASSELL et al. (1976) and HASSELL (1978) for a comprehensive review of these studies).

5.2 The prey death rate—the effect of prey density

SOLOMON (1949) defined a *functional response* as a change in the numbers of prey attacked in a fixed period of time by a single predator when the initial prey density is changed. Intuitively we expect such a response to be an increasing function of prey density and that there must be some limit beyond which a predator cannot increase its attack rate further. This idea was elaborated by HOLLING (1959a) who suggested that functional responses could be divided into three categories (Types I, II and III in Fig. 5–1). These responses are best understood from simple models.

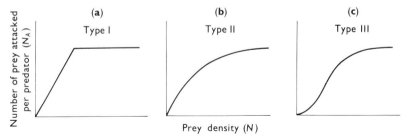

Fig. 5–1   Three types of functional response showing the changes in the number of prey attacked per unit time by a single predator as the initial prey density is varied. (After HOLLING, 1959a.)

To obtain a 'type I' response (Fig. 5–1a), we need merely assume a constant rate of encountering prey (the attack coefficient a') and a threshold prey density (N_x) above which there is no further feeding by the predator. This situation corresponds to the behaviour of some filter-feeders; prey intake is proportional to prey density until the predator is satiated when it abruptly ceases feeding. Thus

$$N_A = a' T_S N \qquad \text{when } N < N_X \qquad (5.1a)$$

and

$$N_A = a' T_S N_X \qquad \text{when } N > N_X \qquad (5.1b)$$

where N_A is the number of prey attacked per predator, T_S is the time available for searching (a constant in this case) and N is the number of prey available. For the sake of simplicity, we have assumed the prey to be constantly replenished as eaten so that the value of N does not decline during the time period, T_S (see below). The constant, a', will appear repeatedly in this and the next chapter, and we should be quite clear that it is an instantaneous rate of discovering prey by one predator—the N_A/N per unit of searching time as seen by rearranging equation (5.1a). Its value depends to a large extent on the predator's activity and ability to perceive prey. Equations (5.1), therefore, predict a linear relationship (of slope $a' T_S$) between the number of prey encountered (N_A) and prey density (N) for each predator, up to a certain level, N_X, when it abruptly levels off due to satiation.

The 'type II' response as shown in Fig. 5–1b and from the examples in Fig. 5–2, is widespread amongst insect parasitoids and invertebrate predators generally; the number of attacks per predator now shows a negatively accelerating rise to an upper plateau. Such functional responses were predicted by HOLLING (1959b) from a simple model developed by deductive reasoning. He disputed that the searching time, T_S in equation (5.1), can be constant. The acts of quelling, killing, eating and digesting a prey are time-consuming activities, which he collectively called the '*handling time*', and reduce the time available for further search. Thus,

$$T_S = T - T_H N_A \qquad (5.2)$$

where T is the total time available and T_H is the handling time. By substitution in equation (5.1) we now have the functional response equation:

$$N_A = a'(T - T_H N_A)N$$

or (5.3)

$$N_A = \frac{a'NT}{1 + a'T_H N}$$

This is the 'disc equation' of HOLLING (1959b), so-called because it was supported by experiments where a blindfolded subject (the predator) searched for sandpaper discs (the prey) on a flat surface. The form of the functional responses predicted from this model depends upon two parameters, the attack coefficient, a', and the handling time, T_H. The value of T_H determines the maximum number of prey that can be attacked within the time T (i.e. T/T_H) and a' defines how rapidly the response rises to this level. The curves in Fig. 5–2 are accompanied by estimates of both a' and T_H. The manner in which these are obtained from the data is fully described by ROGERS (1972) and takes into account the depletion of available prey (or unparasitized hosts) as the experiment proceeds. Rearrangement of equation (5.3), as suggested by HOLLING (1959b), to give the linear equation

$$N_A/N = a'T - a'T_H N_A \qquad (5.4)$$

will also enable a' and T_H to be estimated from the slope $(=a'T_H)$ and intercept $(=a'T)$, but this is only appropriate if prey are replenished as eaten (i.e. N remains constant during the time period, T).

It is, of course, an oversimplification for the predator's attack rate to depend solely on two constants. For example, we would expect the handling time to be a variable, often increasing as satiation is approached. Similarly, the parameter a', that depends largely on the activity of the predator and its ability to perceive prey, is likely to vary depending on hunger levels (or egg depletion in parasitoids). In spite of this, the disc equation describes a wide range of experimental results and has been invaluable in focusing on *time* as an important 'resource' for a predator.

The disc equation will be an important component in the development of a general predator–prey model discussed in the next chapter. It permits, in the case of parasitism, prediction of the number of hosts encountered per parasitoid within an area. But for predators where

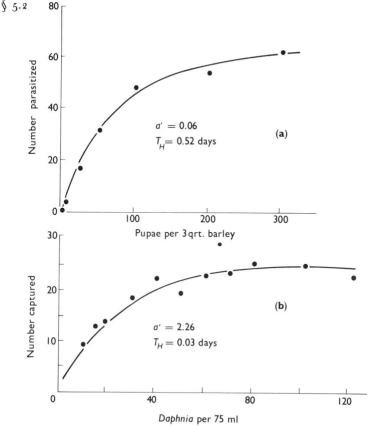

Fig. 5–2 A typical, type II functional response from an insect parasitoid and predator. The attack rate, a', and handling time, T_H, are shown in each case (a) The parasitoid wasp, *Nasonia vitripennis*, parasitizing housefly pupae. (After DEBACH, P. and SMITH, H. S. (1941), *J. econ. Ent.*, **34**, 741–5.) (b) The larva of the damsel fly, *Ischnura elegans*, feeding on the water flea, *Daphnia magna*. (After THOMPSON, D. J. (1975), *J. Anim. Ecol.*, **44**, 907–16.)

juveniles as well as adults must search for prey, the situation is more complex in one important respect. The functional response will change during the development of the predator. We need to know, therefore:

(1) how the parameters a' and T_H vary between successive developmental stages of the predator when they search for the same size of prey and
(2) how they vary within the same predator stage that feeds on different sizes of prey.

Figure 5–3 shows an example where functional responses have been obtained for four different instars of a coccinellid larva feeding on aphids of the same size. In this case, the different responses hinge almost exclusively on a decreasing T_H with successive predator instars. In general, we may expect *both* a decrease in T_H and an increase in a' as predator size increases for a given prey size (Fig. 5–4). Similarly, for any particular predator size there should be an increase in T_H and perhaps also a reduction in a' as the prey get larger (larger prey take longer to eat and are more difficult to catch).

The sigmoid, 'type III' functional responses (Fig. 5–1c) are con-

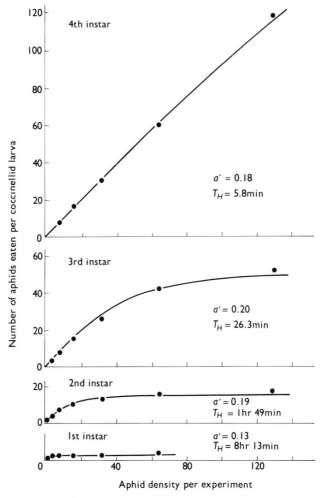

Fig. 5–3 Functional responses of four different larval instars of the coccinellid, *Harmonia axyridis*, feeding on aphids during a 24-h period. Note the considerable differences in handling time between instars. (After MOGI, M. (1969), *Jap. J. appl. Ent. Zool.*, **13**, 9–16.)

ventionally thought to result from learning in vertebrate predators. We have already seen in the last chapter how TINBERGEN's (1960) theory of specific searching images can lead to sigmoid relationships between the per cent of a prey species in a predator's overall diet and the density of that particular prey (Fig. 4–1). This implies that the functional response (N_A plotted against N) will also be sigmoid provided that the total number of prey eaten remains constant. Such a functional response has been most clearly demonstrated by HOLLING (1965). He devised an experiment where a deermouse was presented with various densities of a preferred prey (sawfly pupae buried in sand) and an excess of alternate

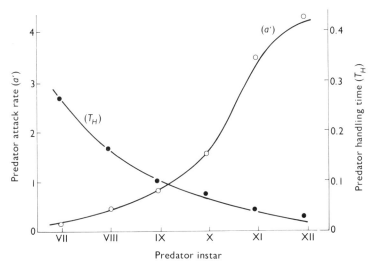

Fig. 5–4   The attack rate, a'(o) and handling time, T_H(●) of different larval instars of the damsel fly, *Lestes sponsa*, feeding on *Daphnia* of a constant size. (By kind permission, D. J. THOMPSON and J. H. LAWTON; unpublished.)

food (dog biscuits in a feeding hopper). The resulting functional response is shown in Fig. 5–5. As sawfly density increases, the deermouse learns to search in the sand for its preferred prey rather than feed on the dog biscuits; hence the increasing proportion of sawflies taken as their density increases (between points A and B in Fig. 5–5). The response begins to level off due to the combined effects of satiation and handling time acting in the same way as in the type II response. These kinds of learning must be widespread amongst vertebrates. The real difference from the previous responses is that the initial time available for search for a particular type of prey, T, is no longer a constant. It increases with prey density, probably as a sigmoid function itself. When prey density is low, the predator will concentrate on other prey or seek alternate areas where prey density may be higher (ROYAMA, 1970; see section 4.3).

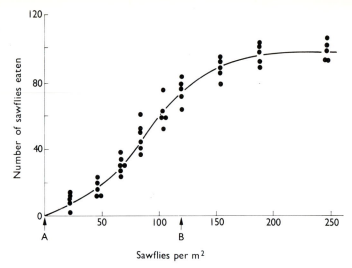

Fig. 5–5 A type III functional response of a deermouse to its sawfly prey when an excess of an alternative food (dog biscuits) is also provided. An increased *proportion* of prey are taken between points A and B. (After HOLLING, 1965.)

5.3 The prey death rate—the effect of predator density

The searching performance of several parasitoids and predators under laboratory conditions has been shown to depend on the density of the searching animals themselves. This is quite clear from the relationships in Fig. 5–6, from parasitoids (a to c) and arthropod predators (d and e). In all cases, predators at the various densities shown have been exposed to a fixed number of prey for the experimental duration, at the end of which the numbers of prey eaten (or hosts parasitized) have been scored. The searching efficiency in these graphs (a) is a simple extension of the instantaneous attack rate $(a'=N_A/NT_S)$ to cover the whole period that predators are searching for prey (T_S). Thus

$$a=a'T_S=N_A/N \qquad (5.5a)$$

and is the area of discovery of NICHOLSON (1933) discussed in the next chapter. However, we have already seen that N_A is the number of attacks per predator if the density of available prey remains constant throughout the experiment. To allow for prey depletion during the experiments, we calculate a using the formula:

$$a=(1/P)\ \log_e\ (N/N_S) \qquad (5.5b)$$

where N_S is the number of prey surviving predation. The derivation of equation (5.5b) is explained in section 6.3.

In most cases where such 'mutual interference' between individuals has been found, the biological mechanism is unknown. However, studies on *Nemeritis canescens*, a parasitoid of stored product moths,

41

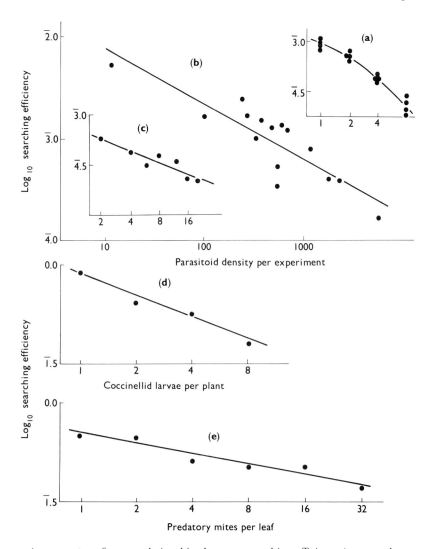

Fig. 5–6 Interference relationships between searching efficiency (expressed as the area of discovery) and parasitoid (**a** to **c**) or predator (**d** and **e**) density per experiment, on logarithmic scales. (**a**) *Pseudeucoila bochei*. (Data from BAKKER, K., BAGCHEE, S. N., VAN ZWET, W. R. and MEELIS, E. (1967), *Entomologia exp. appl.*, **10**, 295–311.) (**b**) *Nemeritis canescens*. (After HASSELL, M. P. and HUFFAKER, C. B. (1969), *Res. Popul. Ecol.*, **11**, 186–210.) (**c**) *Cryptus inornatus*. (Data from ULLYETT, G. C. (1949), *Can. Ent.*, **81**, 285–99.) (**d**) Coccinellid larvae (*Coccinella 7-punctata*). (By kind permission, S, MICHELAKIS; unpublished.) (**e**) Predatory mites (*Phytoseiulius persimilis*). (By kind permission, J. FERNANDO; unpublished.)

have shown the interference (Fig. 5–6b) to be largely the result of the reaction of the searching parasitoids to encounters between themselves or to encounters with a host that has already been parasitized. Following interference there is a tendency for the parasitoids to fly or walk away from that particular area. There is, therefore, a reduction in the time available for search, T_S, and consequently a lowered area of discovery (see equation (5.5a)). Coccinellid beetle larvae (Fig. 5–6d) also react markedly to one another on the same leaf, often resulting in one or both falling to the ground. There are also many examples of threat display and aggressive behaviour in parasitoids directed at any 'intruding' individuals of the same species. In all these cases, the time available for search in an area is reduced by encounters with other predators. The advantage of these kinds of behaviour is in providing a mechanism for redistribution of predators and parasitoids when their density in an area becomes high.

The relationships in Fig. 5–6 (b to e) have been described by the simple model (from HASSELL and VARLEY, 1969).

$$\log a = \log Q - m \, \log P$$

or $$\hspace{6cm} (5.6)$$

$$a = QP^{-m}$$

where m, the mutual interference constant, is the slope of the relationships and log Q, the intercept, is the value of log a when the log density of predators is zero (i.e. when $P = 1$).

While this provides an adequate description of all the data sets in Fig. 5–6, it is not a realistic model over a wide range of predator densities. The searching efficiency cannot continue to rise indefinitely as predator densities get progressively lower. Much more likely is a curvilinear response, levelling off at very low predator densities when interference is negligible (as in Fig. 5–6a). Indeed, models based on the idea that there is a period of inactivity following each encounter between predators, predict just such curvilinear relationships (ROGERS and HASSELL, 1974; BEDDINGTON, 1975).

Many predators exhibit the opposing phenomenon, 'co-operation', which although relatively rare in invertebrates, is well seen in pack- or flock-hunting vertebrates such as wild dogs. In these cases, interference is more likely between the co-operative groups rather than between individuals, and will be behaviourally much more complex than in the simple invertebrate examples in Fig. 5–6.

We have seen in the last two sections that prey and predator densities can affect predation by their influence on T_S, the searching time of a predator. The value of T_S is reduced by both increases in prey density (due to increases in total handling time) and increases in predator density (due to increased interference). Searching time is also important in the next section, in which the way predators may apportion their searching time in areas of different prey density is discussed.

5.4   The prey death rate—the effect of prey distribution

We shall see in the next chapter that most theories for the interaction of predator and prey populations assume that the predator searches at random for prey. Each prey individual, therefore, has an equal probability of being discovered. In the real world, however, the prey population is not a single homogeneous unit; it has a spatial distribution (between leaves, branches, trees, etc.) that is usually clumped. Some unit areas have a high prey density while others a low one. Within this framework, random search implies that, on average, the same time is spent per predator in each unit area; or, equivalently, that the same number of predators are searching in each.

While random search may be a convenient assumption mathematically, it is certainly not a realistic one biologically. A prudent predator will tend to spend more time searching in those areas where the prey are plentiful rather than scarce. Some examples where such aggregation has been quantified are shown in Fig. 5–7, coming from birds and insects. Despite the range of forms to the relationships (sigmoid, convex and concave), it is reasonable to suggest that the idealized response to prey distribution will tend to be sigmoid as shown in Fig. 5–7a. There is a lower plateau where the predators do not distinguish between different low prey density areas, an upper plateau where they do not distinguish between the high density areas, all of which yield ample food, and an intermediate zone where the predators discriminate markedly between different prey densities. Although only Fig. 5–7a, from a field study, approaches this ideal, the other examples, from laboratory experiments, may be sensibly interpreted as lying on different regions of this complete response.

There are many types of behaviour that will lead to a predator remaining longer, on average, in areas of relatively high prey density. The range is well shown by predatory arthropods. There may be a 'long-distance' attraction to some factor whose concentration is a function of prey density. The parasitoid wasp, *Nemeritis*, for example, is strongly attracted by the crowding pheromone from its host, a moth living in cereals (CORBETT, 1971), and several species of insect predators and parasitoids are attracted by the aggregation and sex pheromones released from their bark beetle prey. Alternatively, or in addition, many species 'sample' different prey areas, remaining in any one while the rate of prey capture is above some threshold level, probably defined by the degree of hunger. Several species of insects and mites achieve the same end by simple behavioural changes after feeding. They show an increased turning rate and perhaps a reduced speed of movement after feeding upon a prey, which leads to more intensive search in that immediate area. This behaviour then tends to decay to the prefeeding pattern if not reinforced by further location of prey.

Whenever predators aggregate where prey are abundant, the likelihood of mutual interference is enhanced. It is in this light that we

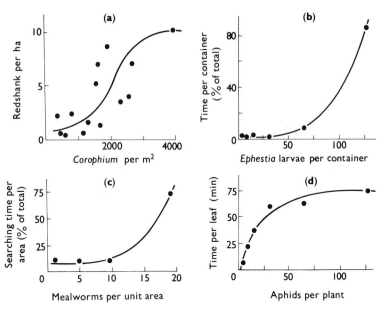

Fig. 5–7 Aggregative responses by predators and insect parasitoids. (a) Redshank (*Tringa totanus*) searching for their amphipod prey on a sea shore. (Data from GOSS-CUSTARD, J. D. (1970), *J. Anim. Ecol.*, **39**, 91–113.) (b) The parasitoid wasp, *Nemeritis canescens*, searching for flour moth larvae in a cage. (From HASSELL, M. P. (1971), *J. Anim. Ecol.*, **40**, 473–86.) (c) Great tits (*Parus major*) searching for mealworms in a small room. (After SMITH, J. N. M. and DAWKINS, R. (1971), *Anim. Behav.*, **19**, 695–706.) (d) The parasitoid, *Diaeretiella rapae*, searching for aphids on a plant. (By kind permission, T. A. AKINLOSOTU; unpublished.) (After HASSELL and MAY (1974); courtesy of Blackwell Scientific Publications.)

should view the selective advantage of interference; as a mechanism increasing the likelihood of dispersal from areas where prey are already, or are likely to become, heavily exploited. The dispersing predators then have the chance of locating further less-exploited areas of prey. This dynamic balance between the two processes of aggregation and interference should thus act to leave an optimum number of predators within a given area. It will assume especial importance for such parasitoids as *Nemeritis* (Figs. 5–6a and 5–7b) which continue to be attracted to areas of hosts irrespective of whether or not the hosts are parasitized.

5.5 The predator rate of increase

An insect parasitoid is a special, and simple, case of predator whose rate of increase, or numerical response, is usually a linear function of the number of hosts parasitized—each host parasitized tends to contribute a fixed number of parasitoid progeny to the next generation. It is to this assumption that the so-called 'predator–prey' models in the next chapter adhere. For true predators, however, such a linear relationship between prey eaten and predator rate of increase is a great over-simplification (BEDDINGTON *et al.*, 1976). Each successive developmental stage must usually find and eat several prey items in order to survive. In the case of adult females, there will also be a dependence between prey eaten and fecundity. The overall rate of increase of the predator population will therefore depend on (1) the survival rate of each developmental stage and (2) the fecundity of the adults, both of which depend importantly on the rate at which the predators consume prey during the course of their development.

The rate of increase of a predator population can be markedly reduced by poor survival of the immature stages even if the adult female fecundity is high. The relationship between predator survival and the density of available prey is therefore an important one and found to be relatively uniform, at least amongst arthropod predators for which there is most information. The proportionate survival is zero at very low prey densities (they all starve), and then increases with prey density up to a plateau where the two become independent. Each predator can now secure sufficient food to complete its development. Figure 5–8 illustrates this for coccinellid larvae and several other examples are given in BEDDINGTON *et al.* (1976).

Once a predator reaches reproductive maturity as an adult, the predator rate of increase depends mainly on fecundity—the number of progeny produced per adult female. We have seen that in parasitoids, this follows directly from the number of hosts parasitized. The relationship also appears simple in many invertebrate predators (e.g. Fig. 5–9) except that there is now a threshold number of prey eaten below which no progeny are produced at all. This merely reflects that some food must be allocated to predator maintenance before there can be any egg

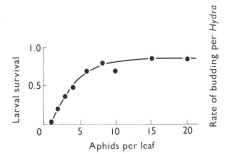
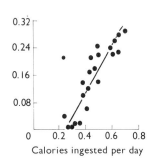

Fig. 5–8 (left)   The proportionate survival of larvae of the coccinellid, *Adalia 10-punctata*, when supplied with different aphid densities. (After DIXON, A. F. C. (1959), *J. Anim. Ecol.*, **28**, 259–81.)

Fig. 5–9 (right)   The reproductive rate of *Hydra pseudoligactis* as a function of the amount of food eaten. (After SCHROEDER, L. (1969), *Ecology*, **50**, 81–6.)

production. Although not clearly shown from this example, such relationships must tend to level off at high levels of prey ingestion when the maximum fecundity per female is achieved. However, by being linear over a considerable range of prey intake, it follows that the relationship between fecundity and prey *density*, rather than prey *eaten*, should take a similar form to the appropriate functional response to prey density for that predator. This is indeed the case, as shown in Fig. 5–10 for coccinellid beetles and predatory mites—the relationships correspond to type II functional responses except in being 'displaced' along the prey density axis, reflecting the minimum prey density for any egg production.

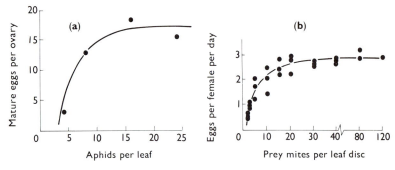

Fig. 5–10   The fecundity of two predatory arthropods as functions of the density of prey available. (**a**) The coccinellid, *Adalia 10-punctata* feeding on aphids. (After DIXON, 1959.) (**b**) The mite, *Phytoseiulus persimilis*, feeding on prey mites. (After PRUSZYNSKI, S. (1973), *SROP/WPRS Bull.*, Integrated Control in Glasshouses, 41–46.)

The components discussed in this chapter provide the basis for a general predator–prey model in which the dynamic effects of each component can be explored. Such a model has yet to be developed. The tendency to date has been to concentrate on much simpler models which are more appropriate to insect parasitoids than to other predators. They do not include the important age class effects found in predator populations, such as the parameters, a' and T_H, scaling with the stage of predator development. Nor do they include the dependence of the predator rate of increase on prey density as discussed in this section. On the other hand, several of the components of searching behaviour, such as the effects of handling time, mutual interference and aggregation, *have* been included in theoretical models, and these models have led to an understanding of how each component is likely to affect the fluctuations in predator and prey populations. Some of these models are discussed in Chapter 6.

6 Theories of Predator–Prey Interactions

6.1 Some experimental models

This chapter commences with a discussion of experiments which illustrate the kinds of population fluctuations of predators and prey that theoretical models seek to predict and explain. Laboratory experiments are used since it is so difficult under natural conditions to disentangle the part played by predation from all the other complicating factors affecting the prey populations; factors that all play a part in determining population levels and fluctuations (SOLOMON, 1969). Predators, too, are often polyphagous, act as food for other predators, compete amongst themselves or succumb to disease. The problem in natural environments, therefore, is essentially a multispecies one, and very difficult to analyse. We would be wise in the first place to understand the much simpler two-species interactions from laboratory experiments.

The first detailed experiments with predators and prey were devised by GAUSE (1934) who cultured together two ciliate protozoan populations, *Paramecium caudatum* and its predator, *Didinium nasutum*, and sampled their population size at regular intervals. Both species were cultured in small, water-filled vessels with a medium consisting of a filtered extract of oatmeal upon which were feeding a variety of bacteria that served as food for the *Paramecium*. Gause found that the outcome of his experiments differed markedly under different conditions. In the simplest cases, the predators would normally exterminate the prey and then starve, irrespective of the initial densities of both populations (Fig. 6–1a). Gause was able to reverse this outcome by introducing a *prey refuge* into the system. An oat medium containing a sediment was added to the water and this served as a refuge for the prey into which *Didinium* would not enter. Now the experiments led to the extinction of *Didinium* after all the exposed *Paramecium* had been eaten. The *Paramecium* population then increased rapidly from the stock within the refuge (Fig. 6–1b).

These experiments tell us little except that interactions may generally be rather unstable (one species becoming extinct) within such small uniform systems. In a third series of experiments, Gause regularly introduced 'immigrants' (one *Didinium* and one *Paramecium*) into the vessel at three-day intervals. Obvious oscillations were now observed in both populations with the predator lagging behind the prey population by approximately two days (Fig. 6–1c). These experiments show that predation can generate repeated oscillations in predator and prey populations, although Gause believed that they depend upon external factors such as regular immigration rather than being a fundamental

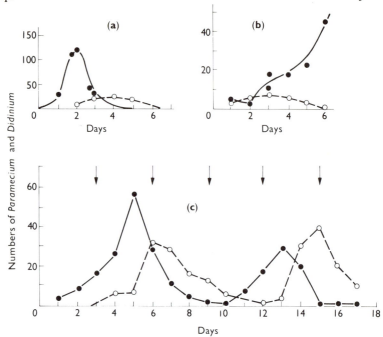

Fig. 6–1 Predator–prey interactions between the protozoans, *Paramecium caudatum* (●) and its predator, *Didinium nasutum* (o). (a) No refuge. (b) With a refuge. (c) With immigration every third day. (After GAUSE, 1934.)

property of the interaction itself. We shall see that the tendency to produce oscillations is fundamental to most single predator–single prey interactions. For this reason, they are commonly seen in laboratory experiments, are certainly a property of all general models and have been suspected in some populations in the field (Chapter 7).

One of the finest examples of sustained predator–prey oscillations in the laboratory comes from some classic experiments by HUFFAKER (1958) using two species of mites; a predatory mite (*Typhlodromus occidentalis*), and its prey mite (*Eotetranychus sexmaculatus*) which were fed on oranges. Like Gause, Huffaker commenced with a relatively simple system in which a variable number of oranges were dispersed amongst rubber balls on a tray, but now there was the added complexity imposed by the spatial arrangment of the prey food. Each orange was partially covered with paraffin wax to present a constant area upon which the prey could feed. Initially, the prey were introduced on their own, serving as a control situation to show the population growth of the prey in the absence of predation (Fig. 6–2a). This conforms very roughly with the predictions of the single-species growth models discussed in Chapter 2 where the populations oscillate about the equilibrium (Fig. 2–8c). When predators were also introduced, the outcome was similar to Gause's first

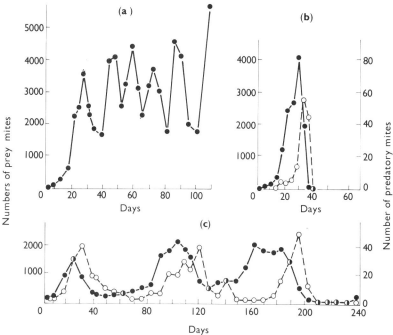

Fig. 6–2 Predator–prey interactions between the mites, *Eotetranychus sexmaculatus* (●) and its predator, *Typhlodromus occidentalis* (o) (a) Population fluctuations of *Eotetranychus* without its predator. (b) A single oscillation of predator and prey in a simple system. (c) Sustained oscillations in a more complex system. (After HUFFAKER, 1958.)

experiment; a single oscillation usually followed by the virtual extinction of prey and predator (Fig. 6–2b). The complexity of the system was now increased by employing 120 oranges, each with only one-twentieth of its area exposed. The rapid dispersal of mites from orange to orange was impeded by placing partial barriers of vaseline between oranges. On the other hand, dispersal was facilitated by small upright sticks being placed at intervals throughout the system from which mites could launch themselves on silken strands carried by air currents. The results from such an experiment are shown in Fig. 6–2c. It is clear that the complexity of this system in the form of spatial heterogeneity has conferred considerable stability to the interacting populations.

An even more striking example of persistence of predators and prey in the laboratory comes from UTIDA's (1957) experiments with the Azuki bean weevil (*Callosobruchus chinensis*) and one of its larval parasitoids, *Neocatolaccus mamezophagus*. Utida was able to maintain the weevil–parasitoid interaction within small dishes for over 110 generations, with the only manipulation being the regular replacement of beans at the time when the populations were being sampled.

These experiments show that, while the simplest interactions may be unstable, this is less likely when the prey habitat is made more complex. Furthermore, they suggest that some components of a predator–prey interaction will contribute to the stability of the populations. Identifying these components will help to account for the relative stability of most predator–prey interactions under natural conditions. First we shall consider two of the best known theoretical models and then proceed to show the effects of including more realistic features of predator behaviour.

6.2   The Lotka–Volterra model

This model was independently proposed by LOTKA (1925) and VOLTERRA (1926) and is normally expressed as a pair of differential equations:

$$\frac{dN}{dt} = (r - C_1 P)N$$

$$\frac{dP}{dt} = (C_2 N - d)P$$

(6.1)

where N and P are the prey and predator populations respectively, r is the intrinsic rate of increase of the prey (see page 9) and d is the death rate of predators in the absence of prey. The constant, C_1, may be interpreted as a coefficient of attack, comparable to a' in the previous chapter, and C_2 as a conversion factor of prey into more predator individuals. By framing the model in differential equations, reproduction of both populations is assumed continuous and implies a complete overlapping of generations and continual, non-seasonal breeding. The model, therefore, includes none of the time lags between cause and effect that were discussed in Chapters 2 and 3.

The Lotka–Volterra model is characterized by a particular pattern of population change over time. The populations smoothly oscillate out of phase with each other, with successive oscillations being of the same amplitude (Fig. 6–3a). Furthermore, the amplitude of the oscillations depends upon the initial densities of the predator and prey. Figure 6–3b illustrates this from a graph of the numbers of predators plotted against the corresponding number of prey at any given time (a plot in phase space, cf. Fig. 3–3). The two ellipses come from the same model but with different starting populations, and each represents a single oscillation that the populations will follow indefinitely. The model, therefore, has the biologically unrealistic feature of always 'remembering' its starting conditions. It is said to be *neutrally stable* and is akin to a frictionless pendulum where the amplitude of swing depends on the initial displacement from the point of rest.

The Lotka–Volterra model clearly includes little of the predator behaviour discussed in Chapter 5. The functional responses are linear (as

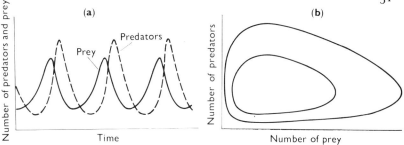

Fig. 6–3 (a) Predator–prey oscillations from the Lotka–Volterra model (*6.1*). (b) The outcome from the model expressed as numbers of predators plotted against numbers of prey. The two ellipses are obtained from different initial population sizes. (After VOLTERRA, 1926.)

type I, Fig. 5–1a, but without the upper plateau), search is at random and the predator rate of increase is directly dependent on the number of prey eaten, as for parasitoids rather than predators.

6.3 The Nicholson–Bailey Model

In 1933, NICHOLSON produced a general theory for the interaction of parasitoids and their hosts, followed in 1935 with a more formal mathematical treatment (NICHOLSON and BAILEY, 1935). Although, as we shall see, the model is more appropriate to parasitoids than predators, it is as much a 'predator–prey' model as that of Lotka and Volterra with which it shares several assumptions. It is, however, formulated in a quite different way. We explore it here in some detail since the remaining models in this chapter are all of a similar, albeit elaborated, form.

Nicholson made three basic assumptions upon which his whole theory rests.

1. Parasitoids search for hosts at random; their behaviour is unaffected by the density and distribution of hosts or other parasitoids in the population.
2. Parasitoids are not limited by their egg supply. Each host found can therefore be parasitized and, similarly, predators never become satiated.
3. Parasitoids of a given species have a characteristic searching efficiency, which Nicholson called the *area of discovery* (*a*, see equations 5.5). This may be thought of as the probability of encountering any particular host in the searching lifetime of the parasitoid. In more physical terms it is the proportion of the total area that is searched by a single parasitoid.

On the basis of these assumptions, Nicholson produced his 'competition curve' that relates the percentage parasitism of the host population to the number of parasitoids searching (A in Fig. 6–4). The

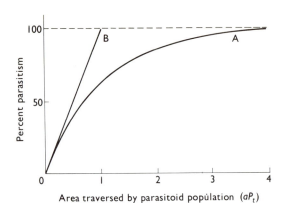

Fig. 6–4 A: Nicholson's competition curve. B: Relationship expected from a systematically searching predator or parasitoid.

decreasing slope of the relationship towards an upper asymptote is a consequence of random search and merely reflects the increasing overlap in searching 'paths' as the parasitoid population increases. Contrast this with a population searching systematically so that the same area is never searched twice. The percentage parasitism would now rise linearly reaching 100% when the parasitoids are numerous enough to cover completely the whole area in their search (B in Fig. 6–4).

The competition curve is the basis of all Nicholson–Bailey models. Given an initial parasitoid population (P_t) and area of discovery (a), the percentage parasitism may be predicted. The number of hosts so parasitized (N_{HA}) becomes the parasitoid population that searches in the next generation (P_{t+1}). This assumes that only a single parasitoid larva can develop within each host attacked and that there is no mortality of the parasitoid progeny. The surviving hosts $(N_s = N_t - N_{HA})$ reproduce with a fixed rate of increase (λ) to give the hosts of the next generation (N_{t+1}). In more formal terms we can express this for the host population as

$$N_{t+1} = \lambda N_s = \lambda N_t \exp(-aP_t) \qquad (6.2a)$$

and for the parasite population as

$$P_{t+1} = N_{HA} = N_t [1 - \exp(-a P_t)] \qquad (6.2b)$$

These equations are based on the first term of the Poisson distribution (PARKER, 1973) which serves to distribute the encounters with hosts randomly to yields the N_{HA} hosts actually parasitized (some will have been encountered more than once). The area of discovery (a) is easily calculated from any set of data provided that the number of parasitoids (or predators) (P_t), the initial density of hosts (N_t) and the number of surviving hosts (N_s) are known. Thus

$$a = \frac{1}{P_t} \log_e \frac{N_t}{N_s} \qquad (6.3)$$

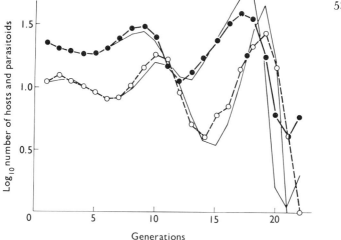

Fig. 6–5 Population fluctuations from an interaction between the greenhouse whitefly, *Trialeurodes vaporariorum* (●), and its parasitoid wasp, *Encarsia formosa* (○). Thin lines show results of a Nicholson–Bailey model where $a = 0.068$ and $\lambda = 2$. (After BURNETT, 1958.)

This equation is derived by replotting the competition curve in Fig. 6–4 in terms of $\log_e (N_t/N_s)$ against P_t, when a linear relationship with a slope of a is obtained that rearranges to give equation (6.3).

The Nicholson–Bailey model is thus a difference equation model implying completely discrete generations. This is its main divergence from the Lotka–Volterra model, and results in quite different stability properties. For each value given to the area of discovery (a) and host rate of increase (λ), there is a unique combination of host and parasitoid densities at which the model is perfectly stable. These equilibrium levels are raised by reducing the value of a or increasing that of λ. It is, however, an *unstable equilibrium* so that the slightest departure of either population from its equilibrium level leads to oscillations of increasing amplitude in both populations. We have, therefore, results of the type seen in the simplest experiments discussed at the start of this chapter; unstable oscillations in the interacting populations. They arise due to the implicit time-lags introduced by the discrete generations. Were we to re-formulate the Lotka–Volterra model as a discrete, difference model, this too would exhibit expanding oscillations.

An example of such unstable oscillations which are also well described by a Nicholson–Bailey model, comes from BURNETT's (1958) experiments with the greenhouse whitefly (*Trialeurodes vaporariorum*) and its chalcid parasitoid, *Encarsia formosa* (Fig. 6–5). These insects were allowed to interact for 23 generations, but the generations were made artificially discrete by limiting each to 48 hours, after which the surviving hosts were doubled and the parasitized hosts assumed to become the adult parasitoids for the next 'generation'. Burnett, in analysing his results, first calculated the area of discovery of his parasitoids for each of the 23 generations of the experiment. He then substituted the average value for a and the artificially fixed value for the host rate of increase ($\lambda = 2$) into equations (6.2a and b). By ensuring that the 'generations' were completely discrete, that the host rate of increase was constant and that each parasitized host gave rise to a single searching parasitoid in the next generation, BURNETT was forcing his system to be of the form of these equations. The only point of uncertainty was whether the outcome of parasitoid search would be adequately represented by a single constant, a. The fairly close agreement between observed and calculated populations suggests that the parasitoids were indeed searching in the way envisaged by Nicholson.

The outcome of Nicholson's models is, however, quite at variance with Huffaker's experiments shown in Fig. 6–2c, and certainly does not help us to account for the relative stability of natural interactions. But still, we should not condemn Nicholson's view of predator and parasitoid population behaviour merely because the models are unstable and natural interactions are not. The models may be made quite stable simply by assuming the prey to be resource limited (competing for food for example) so that the effective rate of increase λ, decreases with increases in prey density (as shown by Huffaker's mites in Fig. 6–2a). This, therefore, would be to imply that predation is destabilizing and that predator–prey interactions persist due to the stabilizing effects of other factors such as competition. We shall now proceed in the next section to show that predation alone *can* stabilize interactions and that the principal weakness of the Nicholson–Bailey model lies in its oversimplification of predator searching behaviour represented by a single constant, the area of discovery.

6.4 Searching efficiency and prey density

Both the Lotka–Volterra and Nicholson–Bailey models make the unrealistic assumption that the number of prey encountered per predator (assuming no exploitation) is a linear function of prey density; that is, type I functional responses (Fig. 5–1a) but without any upper limit. Such relationships, however, apart from implying a potentially infinite appetite (or unlimited egg supply in parasitoids), also deny the importance of handling time, which in itself will change the relationship to the typical invertebrate (type II) functional responses of Figs. 5–1b and 5–2. The effect of this refinement on the outcome of an interaction can be demonstrated by modifying the Nicholson–Bailey model (6.2) to include the additional parameter for handling time, T_H. We have seen from Chapter 5 that high prey densities lead to more prey attacked, more time thus spent handling and less time available for searching. Furthermore, any changes in this searching time, T_S, are directly related to changes in the area of discovery, a, from the equations

$$a = a'T_S = N_A/N = a'T/(1 + a'T_H N) \qquad (6.4)$$

This now gives the population model

$$N_{t+1} = \lambda N_t \, \exp \left(- \frac{a'TP_t}{1 + a'T_H N_t} \right)$$

$$P_{t+1} = N_t \, [1 - \exp \left(- \frac{a'TP_t}{1 + a'T_H N_t} \right)] \qquad (6.5)$$

where the term aP_t from equation (6.2) has been replaced by the more complex expression for a type II functional response. The Nicholson–Bailey model is now only a special case achieved when $T_H = 0$. The inclusion of a finite handling time *always* results in a less stable interaction than the equivalent Nicholson–Bailey model, obtained when $T_H = 0$. Once again, therefore, there is a rapid increase in the amplitude of the population oscillations. This is only to be expected since a type II functional response is one in which a decreasing *proportion* of prey are attacked per predator as prey density increases while the Nicholson–Bailey model assumes a constant proportion. For most predators this destabilizing effect is small since the size of T_H relative to the total time, T, is also small $(T_H/T \ll 1)$. In other words, the periods of non-searching activity associated with eating one prey are small relative to the predator longevity. Model (6.5) therefore includes a general feature of the performance of invertebrate predators, but not one that helps to account for the observed stability of natural interactions. On the other hand, sigmoid, type III functional responses (Figs. 5–1c and 5–5) can certainly contribute to the stability of the interacting populations. The requirement in this case is that the average prey density should lie where the response takes a concave form (between A and B in Fig. 5–5), indicating an increased proportion of prey taken as prey density increases.

6.5   Searching efficiency and predator density

We have seen in section 5.3 and Fig. 5–6 that a decline in searching efficiency per predator with increases in predator density is a frequent feature of laboratory experiments with invertebrate predators and parasitoids. It is a feature that can have a marked effect on the stability of the interaction. To show this, let us substitute equation (5.6) on page 42 into the Nicholson–Bailey model (6.2) by replacing a with the interference expression QP^{-m}, to give

$$N_{t+1} = \lambda N_t \, \exp(-QP_t^{1-m})$$

$$P_{t+1} = N_t[1 - \exp(-QP_t^{1-m})] \qquad (6.6)$$

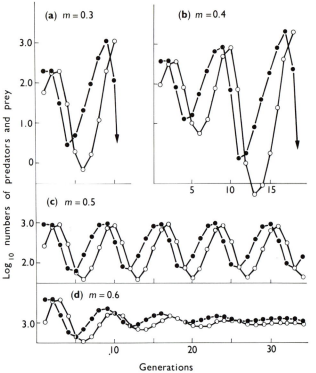

Fig. 6–6 Predator (o) and prey (●) oscillations from model (6.5) showing the progressive stability as the interference constant (*m*) increases from 0.3 to 0.6, with $\lambda = 5$ and $Q = 0.1$. (From HASSELL and VARLEY, 1969; courtesy of Macmillan Journals.)

The Nicholson–Bailey model is now a special case of this model when $m = 0$ (i.e. when there is no interference). No longer are the interactions necessarily unstable: they become progressively more stable as the value of *m* increases. Figure 6–6 illustrates this for a series of models where *m* is increased from 0.3 (unstable oscillations) to 0.6 (damped oscillations). The only other parameter to affect stability is the prey rate of increase, λ, which tends to decrease stability as its value increases. (This is true in almost all population models.) Thus the model in Fig. 6–6b would become stable if λ were reduced from five-fold to two-fold. These stability properties are best seen from the *stability boundaries* in Fig. 6–7, obtained by plotting *m* against λ, which shows the range of values that give unstable and stable interactions. This analysis shows that increasing values of *m* (up to $m = 1$) enhance stability while increasing values of λ oppose this. It should be noted that these values of λ do not necessarily represent the prey fecundity, but are the rates of increase after allowance is made for any mortalities other than predation. For example, a value of $\lambda = 2$ could

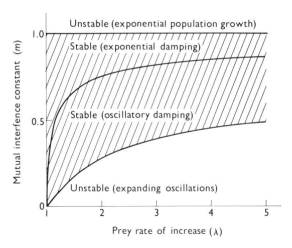

Fig. 6–7 Stability boundaries between the interference constant (*m*) and the prey rate of increase (*λ*) from the model (6.6). The shaded area shows the combinations of *m* and *λ* giving stable predator–prey interactions. (After HASSELL, M. P. and MAY, R. M. (1973), *J. Anim. Ecol.*, **42**, 693–726. Courtesy of Blackwell Scientific Publications.)

result from a fecundity of 100 per adult and an average mortality, due perhaps to a climatic factor, of 98%; or a fecundity of 20 and a further mortality of 90% and so on. Figure 6–7 also shows that there are certain combinations of *m* and *λ* where the populations oscillate about their equilibrium values but others where they increase or approach their equilibria smoothly. This emphasizes that, while predation normally tends to produce oscillations in the populations, there are some conditions when oscillations do not occur. Very marked interference is one of these.

The remaining parameter, Q, in equation (6.6) has no effect on stability but can markedly affect the equilibrium levels of predator and prey populations. High values of Q will depress the equilibria and vice versa, just as high values of the area of discovery in the Nicholson–Bailey model also reduces equilibrium densities. In general, the equilibrium levels are depressed by all factors that increase prey mortality and raise by all factors enhancing the prey's rate of increase.

From this analysis, we may be confident that interference between searching predators can be a powerful stabilizing mechanism. It does not depend upon a high rate of encounters between the predators but, following an encounter, there should be a significant period of time when the predator is not directly searching for prey. This could correspond, for example, to the interval between leaving a particular prey area due to interference and locating a further prey area.

The models so far discussed assume random search by the predator population. Of course, the chance of interference would be increased were predators to aggregate in certain areas. It is to such aggregation in areas of high prey density that we now turn.

6.6   Searching efficiency and the prey distribution

There is an obvious selective advantage in a predator tending to spend most of its searching time where prey are plentiful. Several examples of this were given in section 5.4, Fig. 5–7. At the population level, this will result in more predators searching at any one time in the areas of high rather than low prey density. This behaviour is important not only because it is widespread, but also because of its marked effect on the stability of a predator–prey interaction. To illustrate this, we shall again take the simple Nicholson–Bailey model, but modify it to include the distribution of prey and predators. Firstly, we let the total prey (N_t) and predators (P_t) in each generation be distributed between x unit areas, which may be leaves, branches, trees, etc., depending on the scale of prey clump-size and the searching behaviour of the predator. Thus, coccinellid beetle larvae tend to respond to the aphid distribution per *leaf*, while great tits at nesting time may discriminate between *trees* on the basis of their caterpillar densities. In each of these unit areas (i) the fraction of total prey is α_i and of total predators is β_i. The model for the prey population now becomes (following HASSELL and MAY (1974)):

$$N_{t+1} = \lambda\, N_t \sum_{i=1}^{i=x} [\alpha_i \exp(-a\, \beta_i\, P_t)] \qquad (6.7)$$

This, therefore, distributes P_t predators and N_t prey into x unit areas in the proportions specified by β_i and α_i. For simplicity, consider the case where there are only five unit areas ($x=5$) containing the following fractions of prey and predators in each generation.

Area	(1)	(2)	(3)	(4)	(5)
Fraction of prey (α_i)	0.35	0.30	0.20	0.10	0.05
Fraction of predators (β_i)	0.53	0.35	0.10	0.01	0.01

Figure 6–8b contrasts this model, which is stable, with the comparable Nicholson–Bailey model in Fig. 6–8a where search is random (i.e. $\beta_i = 0.2$ for each area). Clearly, aggregative behaviour on the part of predators can contribute significantly to the stability of an interaction. It depends also, of course, on the distribution of prey. No amount of aggregation may be sufficient to provide stability if the prey evenly dispersed.

The stabilizing effect of predator aggregation is due to the relative protection of prey in low density areas and their increased susceptibility where abundant. The low prey density areas are therefore partial refuges from predation and act in a similar way to the physical prey refuges

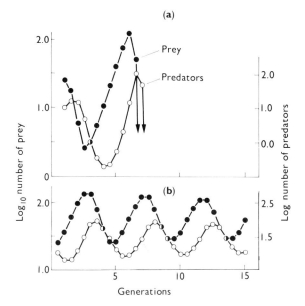

Fig. 6–8 Predator (o) and prey (●) oscillations from model (6.7) showing the stabilizing effect of predator aggregation in regions of high prey density. (a) Nicholson–Bailey model with $a = 0.11$ and $\lambda = 2$. (b) As above, but predators and prey in each generation are distributed between 5 unit areas as specified in the text.

discussed in section 4.3. An absolute refuge for a fixed *proportion* of prey, for example, can stabilize an interaction, but the proportion of protected prey is critical. If too small, the effect on stability is negligible, in the same way as very weak aggregation may be. If too large, too many prey are protected for any significant effect by the predator. Refuges where always the same *number* of prey are inaccessible in each generation are more powerful stabilizing mechanisms, inevitably since a greater proportion will now be protected at low rather than high prey densities.

We have considered in isolation, three ways in which the Nicholson–Bailey model may be made more realistic. The models, however, remain most simple. An obvious next step would be to combine the effects of prey density, predator density and prey distribution in a single model. This is easily done and gives the expected result that interference and aggregation combine to provide a powerful stabilizing mechanism which is opposed by the effects of handling time. A more important progression will be to include in a theoretical model not only the factors affecting the prey death rate, but also those affecting the predator rate of increase outlined in section 5.5. These are likely to have significant effects on the outcome of interactions and the models will then cease to apply strictly to parasitoids alone.

7 The Role of Predators in Natural Ecosystems

7.1 Density dependence in predator–prey interactions

We have seen from the population models in Chapter 6 that the simplest assumptions about predator behaviour give rise to unstable oscillations in a closed predator–prey system. If the prey are resource limited, the interaction can become stable, but stable interactions can also be achieved solely by making predator behaviour more realistic to include, for example, interference and aggregative responses to the prey distribution. In these cases the effect of predation is usually still to induce oscillations whenever the populations are moved from their equilibria, but these gradually dampen and the equilibrium is reapproached.

Since predators can be the cause of a stable equilibrium, we might expect them to act from generation to generation in a density dependent way. In other words, following the definition in Chapter 1, there should be a greater percentage predation in generations when prey density is high than when prey density is low. This, however, is an over-simplification. Let us consider, for example, the experimental interaction in Fig. 6–5, taking the percentage parasitism by *Encarsia* in each generation and plotting this against the corresponding whitefly population, as shown in Fig. 7–1. Such a plot, of course, would reveal any

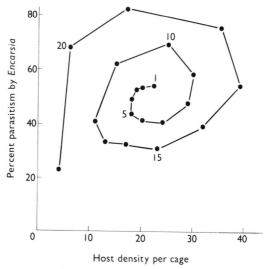

Fig. 7–1   A delayed density dependent relationship obtained from the observed results in Fig. 6–5. (After VARLEY *et al.*, 1973; courtesy of Blackwell Scientific Publications.)

density dependence in the sense defined above, but instead of a positive relationship, there is now a pronounced anti-clockwise spiralling of the points when joined in the sequence of generations. This striking pattern is quite simply the consequence of the two populations oscillating out of phase with each other, in much the same way that plotting the predator population against that of the prey in the Lotka–Volterra model (Fig. 6–3b) yielded closed ellipses. The fact that the spiral is unwinding in Fig. 7–1 merely reflects that the oscillations are increasing in amplitude. Were they to decrease in amplitude towards the equilibria of the populations, the spiral would inevitably 'wind in' towards a stable point.

Because such relationships are so very different in appearance from the density dependent ones envisaged by HOWARD and FISKE (1911) and SMITH (1935), the term *delayed density dependence* was coined by VARLEY (1947) to describe the action of a parasitoid population acting with the time delays inherent in the Nicholson–Bailey model. Although so precisely defined, it remains a useful term whenever such spiralling is detected. We find, however, that there are relatively few examples of delayed density dependence in the ecological literature, which in turn only indicates how rarely predator and prey populations do undergo regular, cyclic oscillations. Only when two necessary conditions have been met are such oscillations likely to be observed.

(1) The predator population must be the *key factor* (VARLEY & GRADWELL, 1970) causing population change (i.e. fluctuations in predation must be largely responsible for fluctuations in the total prey mortality). Without this condition, any tendency for predator-induced oscillations will tend to be obscured by other important factors, due to changes in climate, disease, competition, etc.

(2) The predator population should be more-or-less specific to the particular prey species. Polyphagous predators are much less likely to show numerical changes associated with the abundance of only one of their prey, and hence will not show a typical delayed density dependent relationship.

The general absence of marked predator–prey oscillations under natural conditions therefore indicates how few predators are both specific to a given prey species and also a major cause of fluctuations in the prey population. Even when regularly oscillating populations are observed, it is usually difficult to ascribe this with certainty to predation. The much-debated cycles in the numbers of snowshoe hare and lynx in Canada, deduced from fur returns since the last century, are a case in point. They are unlikely to be true predator–prey oscillations in a causal sense. Other factors (probably related to food supply) seem to be responsible for the hare cycles and the predators are merely 'following' these changes rather than 'driving' them along.

The usual role of predators in natural systems is therefore a less obvious one than seen in laboratory experiments, but still may be a most important one to the observed population dynamics. Let us consider, for example, a polyphagous predator and one of its prey (species X). Because of its wide range of food types, the predator population size does not depend on fluctuations in species X; X is heavily predated when abundant but when rare the predator seeks food elsewhere. Predation, under these conditions, will not tend to produce oscillations in X, but is likely to act as a true density dependent mortality contributing to stability in a similar way to intraspecific competition discussed in Chapter 2. An example of this kind is shown in Fig. 7–2a where the supposed predation per year of

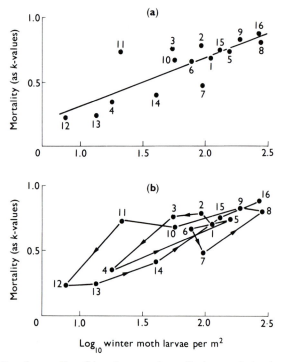

Fig. 7–2 Pupal mortality of the winter moth ascribed to predation (expressed as k-values) plotted against the log winter moth larval density between 1950 and 1965. (a) Linear regression through observed points (b) Anti-clockwise spiralling when consecutive points are linked. (Data from VARLEY, G. C. and GRADWELL, G. R. (1968), *Symp. R. ent. Soc. Lond.*, **4**, 132–42.)

winter moth (*Operophtera brumata*) pupae while in the soil (approximately from June to November), is compared with the density of larvae descending from the trees to pupate. The relationship is clearly density dependent and will contribute to the stability of the winter moth population. It can be explained (as for species X above) by assuming that the predators (mainly carabid and staphylinid beetles and shrews) remain relatively constant in number from year to year and tend to feed heavily on pupae only when pupae are plentiful—perhaps by searching more intensively under the oak trees rather than elsewhere. There is, however, a further component in this example that is apparent from Fig. 7–2b. This differs from Fig. 7–2a only in that the points are now serially linked from 1950 to 1965 (shown as 1–16 respectively) resulting in a pronounced anti-clockwise spiralling as in Fig. 7–1, but now more elliptical. The explanation is not hard to find. One or more of the predator species involved, probably beetles, are showing a greater rate of population increase in years when winter moth pupae are abundant. Their numbers are thus tending to follow the more-or-less cyclic changes in winter moth populations that were observed during this period. The overall picture to emerge, therefore, is one of predation tending to be a rising function of the winter moth population density (the density dependent component—see Fig. 7–2a) which at the same time is slightly obscured by a weak tendency for the predator rate of increase (its numerical response) to depend also on prey density, giving the time-delayed response of Fig. 7–2b.

It is clearly not going to be possible to ascribe a single role to predators in natural communities. There may be a few instances of predator–prey cycles; there are likely to be many cases where predation is a density dependent mortality contributing to stability as in the winter moth example, and there are probably as many or more instances where predation has little effect and is but one of many mortalities affecting a prey population. Unfortunately there are still relatively few studies where the role of predation has been adequately demonstrated under natural conditions. This requires that the percentage predation be known for several generations before there can be any confidence in its observed relationship with prey density. We can be confident, however, that some predators can, and do, maintain their prey at very low levels of abundance in a stable interaction. The many successes in the biological control of pest species by parasitoids, in particular, make this evident.

7.2   Biological control

A successful biological control programme normally follows a predictable sequence of events.

(1) There is an outbreak of the pest species which has been accidentally imported, without its complement of predators and parasitoids (=natural enemies) from its country of origin.

(2) Natural enemies are sought in the country of origin of the pest; in particular, specific natural enemies whose life cycles are preadapted to that of the pest.

(3) The chosen natural enemies are imported, screened under quarantine to ensure that they will not attack beneficial species and then bred in large numbers prior to release.

(4) After release, the natural enemies increase rapidly in numbers, causing a decline in the pest population. This is inevitably soon followed by a reduction in the natural enemy population and pest and natural enemy then persist at very low densities.

The essential part played by natural enemies in such stable interactions is vividly seen from examples where the use of insecticides has dramatically upset the stable interaction between natural enemy and pest. Insecticide application often results in differential and increased mortality of natural enemies compared to that of the prey, which consequently increases rapidly in numbers. Stability at the former low prey population levels may eventually be restored if the pesticide use is discontinued. Detailed examples of this may be found in DEBACH (1974), who also discusses a wide range of successes and failures in biological control.

Examples such as these indicate that at least in insects, specific predators and parasitoids are sometimes responsible for the scarcity of their prey (and hence of themselves). Such low equilibrium levels depend upon two properties of the predator population, as well as on a low effective rate of increase of the prey (see also page 59, Chapter 6).

(1) The predators must have a high effective searching efficiency, a. This requires a high instantaneous attack rate (a' in equation (5.1)) whose effect should not be greatly reduced by the combined effects of handling time, satiation or egg-limitation (T_H in relation to T in equation (5.2)), nor by interference between searching predators (m in equation (5.6)). The value of a will be further enhanced if the predators tend to aggregate in regions of high prey density (β_i in relation to α_i in equation (6.7)).

(2) The predators must not be limited by mortalities during their own life cycles. This would have the effect of reducing their effective rate of increase and so counteracting requirement (1) above (e.g. HASSELL, 1980).

The most vivid demonstration of the importance of (2) above comes from the biological control of the winter moth in Canada (EMBREE, 1965), compared with the winter moth studies by VARLEY and GRADWELL (1970) in Wytham Wood, England. The winter moth was accidentally introduced into Nova Scotia in 1949 and spread rapidly becoming a serious defoliator of hardwood, shade and orchard trees. A biological control programme was initiated in 1954 and two species of larval parasitoids introduced from Europe, a tachinid fly (*Cyzenis albicans*) and an ichneumon wasp (*Agrypon flaveolatum*). Both species became established and increased in numbers. *Cyzenis*, in particular, became very abundant and was the major cause of the decline of the winter moth population, particularly in woodland. The host and both parasitoids still occur in Nova Scotia but at greatly reduced levels, and so the situation has remained for the past ten years. In contrast to this, the winter moth population in Wytham Wood is fairly abundant but rarely defoliates the trees. Furthermore, although *Cyzenis* is present, it has been shown to have a negligible effect on the winter moth population dynamics. The reason for this appears simply to be a difference in the pupal mortalities occurring in the soil in the two countries. In Wytham there is a heavy mortality (between 50% and 95%) that is markedly density dependent, as shown in Fig. 7–2, and, very importantly, which also affects the *Cyzenis* pupae in the soil. Pupal mortality in Nova Scotia, however, is much less severe; in the region of 35%. This difference has been sufficient to lead to the quite different outcomes in the two places (VARLEY *et al.*, 1973; HASSELL, 1980). The strong density dependent pupal predation at Wytham Wood renders the parasitoid ineffective and maintains an average level of winter moth abundance well below the defoliation levels originally seen in Nova Scotia.

The fact that successful cases of biological control have not been followed by violent oscillations as seen in laboratory experiments and predicted from the simplest theories, suggests that 'stabilizing behaviours' such as interference and predator aggregation are important. In particular, the widespread tendency for predators to aggregate where prey are abundant, rather than to search randomly, must be contributing to stability in almost all predator–prey interactions.

7.3 Predators and species diversity

We saw in Chapter 3 that simple competition models have strict requirements for the coexistence of two or more competing species. These criteria can be changed by the inclusion of a predator in the system. In particular, polyphagous predators that apportion their searching time between prey species depending upon their relative abundance, can profoundly increase the range of conditions permitting their prey to coexist. In a single predator and two prey system, for example, such predation can even allow the prey to coexist when the product of their competition coefficients is greater than one ($\alpha\beta > 1$), which would be impossible in the predator's absence (see page 20).

Predation, therefore, can act to increase species diversity by increasing the number of prey species that can coexist. PAINE (1966 and 1974) has provided an excellent example of this from his work on an intertidal community of starfish (*Pisaster*), a carnivorous whelk (*Thais*) and their prey (Fig. 7–3) and four species upon which neither *Pisaster* nor *Thais* feeds. After the deliberate removal of the 'top predator', *Pisaster*, there was a rapid change in the diversity of the system. Within three months,

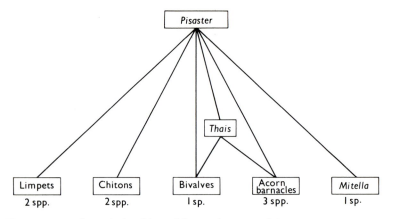

Fig. 7–3 Feeding relationships of the predatory starfish, *Pisaster*. (After PAINE, 1966.)

barnacles (*Balanus*) had occupied from 60 to 80% of the available space. But after one year the mussel, *Mytilus*, and the goose barnacle, *Mitella*, had become dominant and the number of coexisting species was reduced from 15 to 8. At the same time the whelk became very abundant but its limited voracity prevented it from assuming the role of *Pisaster* which can consume from 20 to 60 barnacles at one time. *Pisaster* is consequently able to maintain a much larger community of herbivore species than could occur in its absence, by preventing any competitor from becoming sufficiently abundant to out-compete (for food or space) any of the other species. Its effectiveness is further increased by showing some preference for *Mytilus*, one of the dominant competitors, and by tending to aggregate in regions of high *Mytilus* density. The main ingredients, therefore, for a predator to enhance prey species diversity are that it should be polyphagous, voracious and tend to search non-randomly by feeding most heavily on the abundant prey at any time.

We have seen that the earliest theories of competition and predation were developed in the virtual absence of precise information either from the laboratory or from natural conditions. Their development has had to await the accumulation of such data, which in some aspects is now quite plentiful. For example, there is considerable information on niche overlap and competition in vertebrates, especially in birds (CODY, 1974). There is also much data on the components of predation, mainly from insects (BEDDINGTON *et al.*, 1976; HASSELL *et al.*, 1976; HASSELL, 1980), which should help in the development of future predator–prey models. The choice of experiment or observation is quite critical if it is to relate to theory. Progress will be most rapid when there is a cycle of theory suggesting experiments, that then suggest a modified theory, which lead to further experiments and so on.

References

BEDDINGTON, J. R. (1975). *J. Anim. Ecol.*, **44**, 331–40.
BEDDINGTON, J. R., HASSELL, M. P. and LAWTON, J. H. (1976). *J. Anim. Ecol.*, **45**, 165–86.
BROWN, W. L. and WILSON, E. O. (1956). *Syst. Zool.*, **5**, 49–64.
BURNETT, T. (1958). *Proc. 10th Int. Congr. Ent.*, **2**, 679–86.
CODY, M. L. (1974). *Competition and the Structure of Bird Communities.* Princeton University Press, Princeton.
CONNELL, J. H. (1961). *Ecology*, **42**, 710–23.
CORBETT, S. A. (1971). *Nature*, **232**, 481–4.
CROMBIE, A. C. (1945). *Proc. R. Soc. (B)*, **132**, 362–95.
CROMBIE, A. C. (1946), *Proc. R. Soc. (B)*, **133**, 76–109.
DEBACH, P. (1974). *Biological Control by Natural Enemies.* Cambridge University Press, Cambridge.
DELANY, M. J. (1974). *The Ecology of Small Mammals.* Edward Arnold, London.
EMBREE, D. G. (1965). *Mem. ent. Soc. Can.*, **46**, 1–57
GAUSE, G. F. (1934). *The Struggle for Existence.* Hafner, New York (reprinted 1964).
GIBB, J. A. (1958). *J. Anim. Ecol.*, **27**, 375–96.
GILPIN, M. E. and AYALA, F. J. (1973). *Proc. Nat. Acad. Sci. U.S.A.*, **70**, 3590–3.
GRIFFITHS, K. J. (1969). *Can. Ent.*, **101**, 673–713.
HALDANE, J. B. S. (1949). *Ric. Sci.*, **19** (suppl.), 3–11.
HARDIN, G. (1960). *Science*, **131**, 1292–7.
HASSELL, M. P. (1975). *J. Anim. Ecol.*, **44**, 283–95.
HASSELL, M. P. (1978). *The Dynamics of Arthropod Predator–Prey Systems.* Princeton University Press.
HASSELL, M. P. (1980). *J. Anim. Ecol.*, **49**, 603–8.
HASSELL, M. P., LAWTON, J. H. and BEDDINGTON, J. R. (1976). *J. Anim. Ecol.*, **45**, 135–64.
HASSELL, M. P. and MAY, R. M. (1974). *J. Anim. Ecol.*, **43**, 567–94.
HASSELL, M. P. and VARLEY, G. C. (1969). *Nature, Lond.*, **223**, 1133–7.
HOLLING, C. S. (1959a). *Can. Ent.*, **91**, 293–320.
HOLLING, C. S. (1959b). *Can. Ent.*, **91**, 385–98.
HOLLING, C. S. (1965). *Mem. Ent. Soc. Can.*, **45**, 60 pp.
HOWARD, L. O. and FISKE, W. F. (1911). *Bull. Bur. Ent. U.S. Dep. Agric.*, **91**, 1–312.
HUFFAKER, C. B. (1958). *Hilgardia*, **27**, 343–83.
HUTCHINSON, G. E. (1948). *Ann. N.Y. Acad. Sci.*, **50**, 221–46.
HUTCHINSON, G. E. (1958). *Cold Spring Harb. Symp. Quant. Biol.*, **22**, 415–27.
KENNEDY, J. S. and CRAWLEY, L. (1967). *J. Anim. Ecol.*, **36**, 147–70.
KREBS, C. J. (1972). *Ecology. The Experimental Analysis of Distribution and Abundance.* Harper & Row, New York.
LACK, D. (1945). *J. Anim. Ecol.*, **14**, 12–16.
LACK, D. (1947). *Darwin's Finches.* Cambridge University Press, Cambridge.
LOTKA, A. J. (1925). *Elements of Physical Biology.* Williams & Wilkins, Baltimore.
MACARTHUR, R. H. (1958). *Ecology*, **39**, 599–619.
MACARTHUR, R. H. (1960). *Amer. Nat.*, **94**, 25–36.

MACARTHUR, R. H. (1970). *Theor. Pop Biol.*, **1**, 1–11.

MACARTHUR, R. H. and MACARTHUR, J. (1961). *Ecology*, **42**, 594–8.

MACARTHUR, R. H. and PIANKA, E. R. (1966). *Amer. Nat.*, **100**, 603–9.

MAY, R. M. (1973). *Stability and Complexity in Model Ecosystems.* Princeton University Press, Princeton.

MAY, R. M. (1974). *Theor. Pop. Biol.* **5**, 297–332.

MAY, R. M. (1975). *J. Theor. Biol.*, **51**, 511–24

NICHOLSON, A. J. (1933). *J. Anim. Ecol.*, **2**, 132–78.

NICHOLSON, A. J. (1954). *Aust. J. Zool.*, **2**, 9–65.

NICHOLSON, A. J. and BAILEY, V. A. (1935). *Proc. Zool. Soc. Lond.*, **1935**, 551–98.

PARK, T. (1948). *Ecol. Monogr.*, **18**, 265–308.

PARK, T. (1954). *Physiol. Zool.*, **27**, 177–238.

PARKER, R. E. (1973). *Introductory Statistics for Biology.* Edward Arnold, London.

PAINE, R. T. (1966). *Amer. Nat.*, **100**, 65–75.

PAINE, R. T. (1974). *Oecologia*, **15**, 92–120.

PEARL, R. and REED, L. J. (1920). *Proc. Nat. Acad. Sci.*, *U.S.A.*, **6**, 275–88.

ROGERS, D. J. (1972). *J. Anim. Ecol.*, **41**, 369–83

ROGERS, D. J. and HASSELL, M. P. (1974). *J. Anim. Ecol.*, **43**, 239–53.

ROYAMA, T. (1970). *J. Anim. Ecol.*, **39**, 619–68.

SMITH, H. S. (1935). *J. econ. Ent.*, **28**, 873–98.

SOLOMON, M. E. (1949). *J. Anim. Ecol.*, **18**, 1–35.

SOLOMON, M. E. (1969). *Population Dynamics.* Edward Arnold, London.

SOUTHWOOD, T. R. E. (1962). *Biol. Rev.*, **37**, 171–214.

SOUTHWOOD, T. R. E. (1966). *Ecological Methods with Particular Reference to the Study of Insect Populations.* Methuen, London.

SOUTHWOOD, T. R. E., MAY, R. M., HASSELL, M. P. and CONWAY, G. R. (1974). *Amer. Nat.*, **108**, 791–804.

THOMPSON, W. R. (1924). *Annls. Fac. Sci.*, *Marseille*, **2**, 69–89.

TINBERGEN, L. (1960). *Arch. Néerl. Zool.*, **13**, 266–336.

UTIDA, S. (1957). *Cold Spring Harb. Symp. Quant. Biol.*, **22**, 139–51.

VARLEY, G. C. (1947). *J. Anim. Ecol.*, **16**, 139–87.

VARLEY, G. C. and GRADWELL, G. R. (1970). *Ann. Rev. Ent.*, **15**, 1–24.

VARLEY, G. C., GRADWELL, G. R. and HASSELL, M. P. (1973). *Insect Population Ecology: an Analytical Approach.* Blackwell, Oxford.

VERHULST, P. F. (1838). *Corresp. Math. Phys.*, **10**, 113–21.

VOLTERRA, V. (1926). [Translation in] CHAPMAN, R. N. (1931). *Animal Ecology*, pp. 409–88. McGraw-Hill, New York.

WIT, C. T. DE (1960). *Versl. Landbowk. Onderzoek.*, **66**, 8. p. 82.